MACLEAN, NORMAN
THE DIFFERENTIA

KU-793-421

QH607.M16

WITHDRAWN FROM STOCK
The University of Liverpool

Genetics—Principles and Perspectives: a series of texts

General Editors: Dr. K. R. Lewis, Professor Bernard John

1 The Differentiation of Cells

The Differentiation of Cells

Norman Maclean

Senior Lecturer, Department of Biology, Southampton University

UNIVERSITY LIBRARY LIVERPOOL

LIVERPOOL UNIVERSITY LIBRARY

Edward Arnold

© Norman Maclean 1977

First published 1977
by Edward Arnold (Publishers) Limited
25 Hill Street, London W1X 8LL

Boards edition ISBN: 0 7131 2566 7
Paper edition ISBN: 0 7131 2567 5

All rights reserved. No part of this publication may be
reproduced, stored in a retrieval system or transmitted,
in any form or by any means, electronic, mechanical,
photocopying, recording or otherwise, without the prior
permission of Edward Arnold (Publishers) Limited.

To Margaret, Lorna and Gavin,
with affection

Printed in Great Britain by
William Clowes & Sons, Limited
London, Beccles and Colchester

Preface

This book is intended to provide an introduction to one of the most exciting problems known to man—the differentiation of living cells. How do the diverse types of cells in the same organism, all arising from the same fertilized egg cell, come to be so different? This question, put casually by an unsuspecting lecturer in my undergraduate days, has intrigued me ever since. I hope that this book may help to arouse in some of its readers a similar interest and curiosity.

I have been deeply conscious, while writing the book, of constant oversimplification. Hardly one experiment or observation which has been cited is, on reading the original publication, as definitive or uncomplicated as it has been made to appear. But that is, I suppose, the nature of science. Certainly attempting a synthesis in a general area such as differentiation is somewhat analogous to building a house with bricks all made from different materials from very different localities. That is, pieces of evidence which are put side by side in order to yield a model may be drawn from very dissimilar systems, the one a virus-infected liver cell perhaps, and the other a marine alga. And as I have remarked elsewhere in the text, the areas in which theorizing is easiest but least profitable are those which are very poorly understood. It seems to me that biological science has now moved to a stage where a synthesis of differentiation can be usefully attempted. Thus the book. But since this is a fairly recent development, it would be still very easy to draw inappropriate conclusions, or to use quite irrelevant evidence.

The other difficulty has been a temporal one. I have written this book over a period of two years and much has changed in that time in the field of cellular genetics. Despite occasional revisions, certain parts of the text will be out of date on publication. But most, I trust, will have a slightly longer useful life.

454435

Acknowledgments

I am greatly indebted to past and present colleagues at Southampton, with whom it has been a pleasure to work. Some of them have helped particularly by reading and criticizing parts of the manuscript of this book, especially Dr. Muriel Ord, Dr. David Garrod and Dr. David Morris. Others elsewhere have also provided useful comment and criticism—Dr. David Malcolm, Dr. Godfrey Hewitt and Professor Herbert Macgregor, and particularly the general editors of this series, Dr. K. R. Lewis and Professor Bernard John. To all of these I owe a great debt. I must also thank Mrs. Anne Wharmby for undertaking much of the typing.

Southampton 1976   N.M.

Contents

Introduction: the problem and its importance

There are two biological phenomena which, by reason of their familiarity, are too often taken for granted. Yet they have puzzled and tantalized scientific minds for over a century and we are still far from a proper understanding of either. I refer to the existence of organisms in the discrete categories of species, and the organization of living cells into the distinct groups which we call tissues. The first problem we must leave aside to Darwin and his successors, devoting ourselves in this book to the second problem, an understanding of differentiation.

When a fertilized egg develops into a plant or an animal it does not simply produce a multicellular mass of identical cells. Instead it gives rise to an *organism*, an organized assembly of distinct cell types, the differing cells occurring, for the most part, in discrete tissues. The mystery of the process of differentiation considerably deepened with the realization that differing cells in the same organism all possess an identical and complete set of the genetic material. This discovery ruled out the possibility that differentiation was accomplished by dividing up the genetic material of the fertilized egg into separate portions appropriate to the different tissues. It is clear that, in general, differentiated cells retain a complete set of genes but use only some of them.

We must be aware, however, of the beguiling nature of the theory of selective gene activity in differentiation. For although it affords an understanding of how differentiated cells utilize the genetic information it does not *per se* explain differentiation. As often as not, differential gene activity may be a result of differentiation, rather than its causal mechanism. Moreover, another difficulty is attendant on the view that differentiation is best explained in terms of differential gene activity. It is that in the eukaryotic cell the activity of specific genes can rarely be monitored with any accuracy. This is because transcription of a gene into RNA is not necessarily followed directly by its translation into protein. Indeed, in some cases genes may be transcribed and the RNA never translated. Until such times as different messenger RNA molecules can be accurately partitioned and identified, we must depend on recognition of the specific protein as an indicator of the activity of the individual gene. It should therefore be recognized that present ideas about differential gene activity during cell differentiation, monitored at the level of the protein product, are based largely on inference. Differential gene activity should be therefore seen as an important mechanism in the differentiation process, but not necessarily the primary causal one. How, after all, does a cell 'know' which genes should be expressed and which not?

This leads us to another point which needs to be stressed in relation to cell differentiation. It is that, even in the most complicated eukaryotic organism, there is a very limited number of different cell types. Very often great stress is put on the variety and number of cell types within a multicellular organism. But this should not blind us to what is perhaps even more impressive, namely that in an organism of many billions of cells there is likely to be only scores, or at the most a few hundred *different* cell types. Differentiation is a strictly limited exercise. Within any one cell type there may be many millions of essentially identical cells. Appreciation of this point of eukaryotic organization is surely fundamental to a proper understanding of cell differentiation. It tells us that, whatever mechanisms are involved, they are in many ways analogous to programme selection in a washing machine. The basic differing cell types may each be selected by a fairly simple switch, perhaps a single 'tissue specific master gene'. Once such a switch has been thrown, it will automatically select the appropriate programme of gene expression, turning on and off at appropriate times the many different genes relevant to that cell type. Such a programme selection must often be accomplished long before the cellular characteristics which it determines become overt, and it can clearly persist through many rounds of cell division. Once selected, the programme is normally remarkably stable and confusion with any other programme of gene expression very rare.

Let me stress these two points again because they are, I believe, fundamental to an enlightened view of differentiation. Firstly, differential gene expression, though no doubt important in cell differentiation, does not, of itself, explain how the initial commitment is made. Secondly, this initial commitment need only involve a choice between perhaps a hundred or so separate programmes of gene expression. The striking similarity of different cells of the same differentiated type probably reflects the identical nature of the programme selected within them.

In this book I have endeavoured to give a comprehensive view of the many interacting parameters which combine to induce and maintain cellular differentiation. The relative importance of these different aspects varies from one cell type to another and from one organism to another.

Since one of the easier ways of understanding what differentiation implies is seeing clearly what it does *not* imply, I have chosen to lead in to the problem by way of chapter 1, which is partly devoted to a consideration of cells such as bacteria which are not differentiated from one another, and a look at the earliest symptoms of differentiation in primitive cells and organisms.

1

The evolutionary significance of differentiation

1.1 The origins of differentiation

Differentiation is normally taken to mean the process by which cells and tissues of multicellular organisms become different from one another. Such differences are no doubt one of the necessary consequences of multicellularity, since any organism with thousands or millions of cells is faced with problems of circulation, skeletal support, and movement. None of these difficulties can be overcome without some specialization in the form and function of different parts of the total cell mass.

Even in prokaryotic organization it is possible to recognize some anticipation of the process of eukaryotic differentiation. Thus, the switching on and off of bacterial operons (see p. 70) constitutes an altered commitment on the part of the cell and its metabolic machinery in response to the changing environment, while both division and sporulation demand an even closer commitment to a particular pathway. Some bacteria and blue-green algae form multicellular chains or aggregates, but little or no sign of cellular differentiation is to be found in such colonies. However, one remarkable group of prokaryotes which provides an example of true differentiation is the Myxobacteria. These are small rod-shaped unicellular organisms which form flat spreading colonies on solid media. One subgroup of the Myxobacteria, the fruiting Myxobacteria, will, under appropriate conditions, produce rather tight cell aggregates, within which differentiation occurs to produce large fruiting bodies (Dworkin, 1973). These form as brightly coloured shining droplets which rise above the surface of the cellular mass, and each consists of many spherical cells known as microcytes, a differentiated development from the normal rod shaped cell (Fig. 1.1).

Not only do the myxobacteria provide an example of true differentiation in prokaryotes, but they also form an outstanding example of evolutionary convergence, since the cellular slime moulds or *Acrasieae* engage in a similar process of aggregation and fruiting body production (see p. 7).

Although bacterial differentiation has been cited as an anticipation of eukaryotic differentiation, it seems unlikely that these prokaryotes form part of the main line of biological evolution. An organism like

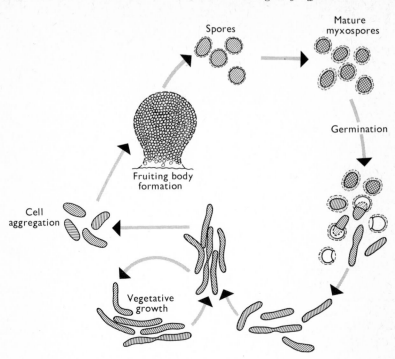

Fig. 1.1 The life cycle of *Myxococcus xanthus*, a member of the Myxobacteria. This figure should be compared to Fig. 1.7, which shows the life cycle of the unrelated organism *Dictyostelium*. (After Fig. 1, from Ashworth, J. M. and Smith, J. E., 1973, Microbial Differentiation. In *Symp. Soc. Gen. Microbiol.*, **13**. Cambridge University Press).

mycoplasma is perhaps a more likely common ancestor of both prokaryotes and eukaryotes. It is then of greater interest to consider the possibility that differentiation within unicellular eukaryotic cells may provide a more accurate appraisal of the origins of the complex pattern of differentiation found in eukaryotes.

For example, three important aspects of differentiation are demonstrated by the Protozoa. The first is a high degree of differentiation and specialization of parts of the single cell. Thus not only do we find mitochondria, chloroplasts, and cortex all possessing considerable independence from nuclear control but, in addition, particular parts of the cortex may be specialized to form sophisticated structures. These include the cilia and flagella which facilitate movement and the oral funnel which serves to channel food into the digestive vacuoles. Some of these features are well illustrated by the large flagellate *Trichonympha*, a symbiotic inhabitant of the gut of termites (Fig. 1.2). We should also notice that this compartmentalization of the Protozoan (and all other

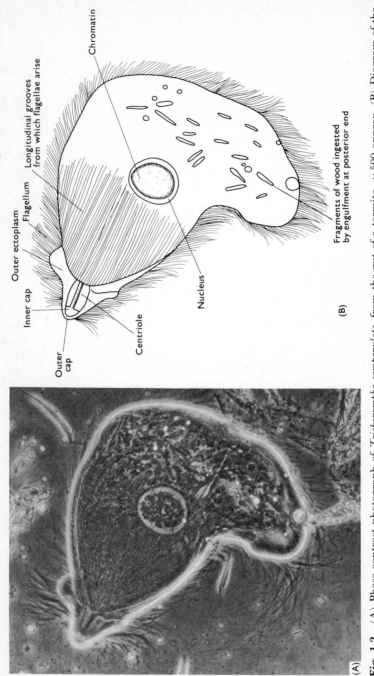

Chromatin

Longitudinal grooves
from which flagellae arise

Flagellum

Outer ectoplasm

Inner cap

Centriole

Outer
cap

Nucleus

Fragments of wood ingested
by engulfment at posterior end

(B)

(A)

Fig. 1.2 (A) Phase contrast photograph of *Trichonympha campanulata*, from the gut of a termite. × 500 approx. (B) Diagram of the photograph shown in 1.2A.

eukaryotic cells) permits mutually incompatible reactions to proceed satisfactorily in different parts of the same cell.

The second aspect of differentiation found amongst Protozoa is the development of syncytial structure, that is the presence of many nuclei in one large acellular cytoplasmic mass. Within unicellular forms a variety of nuclear configurations may be found. Some, like the ciliates *Stentor* (Fig. 1.3) and *Microstomum*, have a highly complex polyploid macronucleus. Others like the large ciliate *Opalina* (Fig. 1.4), found as a common occupant of the frog rectum, have numerous small nuclei. Probably the most impressive multinucleate development in a single celled organism occurs in the life cycle of one group of Myxomycetes known as *Physarum*. The Myxomycetes or slime moulds are eukaryotic organisms related to the fungi. *Physarum* exists as separate amoeboid cells for part of its life history, but many of these amoebae may fuse to form a multinucleate plasmodium containing many thousands of nuclei (Dee, 1962; Sauer, 1973). These nuclei display a high degree of synchrony during DNA synthesis and division (Cummins, 1969). If food becomes scarce, cellular divisions appear within the plasmodium and differentiation into haploid spores takes place. Here we see an experiment in acellular organization, perhaps as an aid to synchronous division. Other syncytial organizations to be found in lower eukaryotes include those of the alga *Vaucheria* and many fungi.

The third aspect of differentiation displayed by Protozoa which merits attention is that of multicellular organization. Examples of this type of organization are rather few, the best occurring in the algae *Volvox* (Fig. 1.5) and *Hydrodictyon*. A fascinating series of algal species show increasingly elaborate multicellular organization, from the basic unicellular *Chlamydomonas* (Fig. 1.6) through *Pandorina* with 16 cells to *Volvox* (Fig. 1.5) with many hundreds. *Hydrodictyon* also displays a large three dimensional structure of very many cells but, like *Volvox*, there is little indication of a division of labour between the cells. Once again, it is amongst the slime moulds that the most interesting example is found. Some slime moulds are grouped under the name of *Acrasiae*. The best known member of this group is *Dictyostelium discoideum*, which, because it displays many characteristics which resemble true differentiation, is widely used as a research material.

Like *Physarum*, *Dictyostelium* normally exists as separate, free living amoeboid cells which wander over the soil surface, feeding on bacteria. When food is scarce, the behaviour of these amoebae changes dramatically. Instead of behaving indifferently to one another, the cells proceed to aggregate into a large tissue mass consisting of thousands of cells. This aggregate appears to result from the release of an attractive compound which has been shown to be cyclic AMP (see p. 138). The aggregate is known as a slug or grex and is capable of co-ordinated movement along light or temperature gradients. It is surrounded by a coating or sheath of slime but the mechanics of movement of the slug

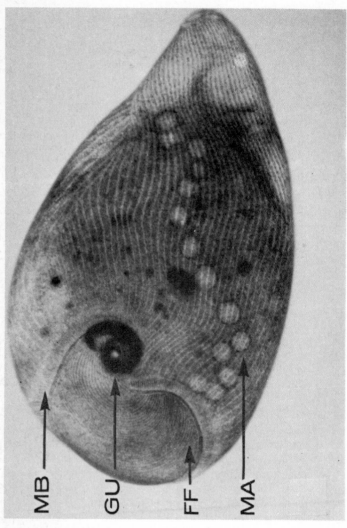

Fig. 1.3   Living *Stentor coeruleus* photographed in a micro-compression chamber. × 190. FF = frontal field; GU = gullet; MB = membranellar band; MA = macronuclear nodes. Numerous food vacuoles are to be seen in the cytoplasm. (After De Terra, N., 1970. *Symp. Soc. Exp. Biol.*, **24**, Copyright Academic Press Inc.)

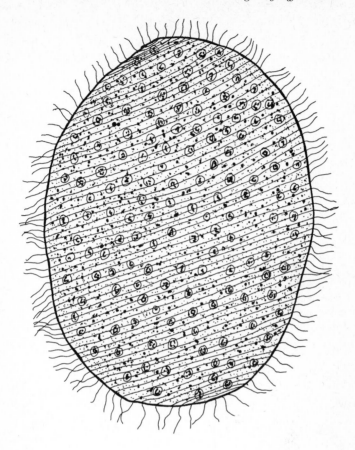

Fig. 1.4 Cell of *Opalina ranarum*, showing numerous nuclei. × 150 approx.

are not understood. Yet a further elaboration on this multicellular organization occurs if food remains scarce. In these circumstances the slug will stop moving, round up, and change into a fruiting body containing many spores (See Fig. 1.7). A remarkable differentiation occurs at this stage, since the stalk cells are those which travelled at the front of the slug, while those cells situated at the rear become spore cells (Ashworth, 1971) (See Fig. 1.8). The striking convergent evolution displayed by this phenomenon and spore production in the Myxobacteria has already been pointed out. In *Dictyostelium* we see perhaps the most impressive exploitation of differential organization shown by any single celled organism. It is also advantageous, as pointed out by

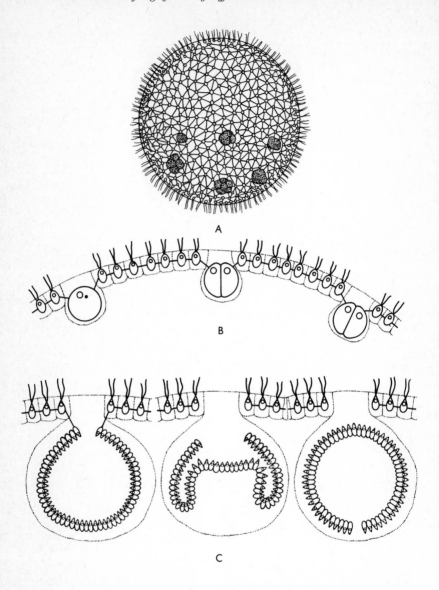

Fig. 1.5   (A) Colony of *Volvox*, with daughter colonies in the lower part of the sphere. (B) Enlarged view of part of a *Volvox* colony showing the cytoplasmic connections between adjacent cells and three daughter colonies commencing growth. (C) A similar section to B above, but showing inversion of the daughter colonies after formation of the definitive cell number. (A and B after Hyman, L., 1940. *The Invertebrates, Protozoa Through Ctenophora*. McGraw-Hill; C after Smith, G. M., 1955. *Crytogamic Botany*, Vol. 1 2nd edn. McGraw-Hill.)

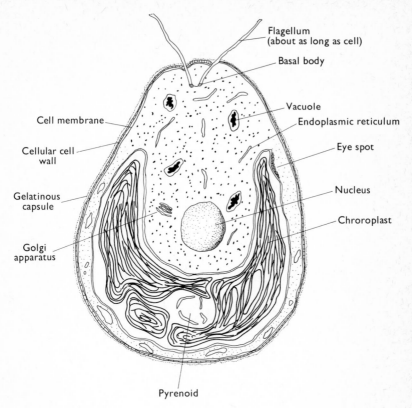

Fig. 1.6 Cell of *Chlamydomonas*. × 10 000 approx.

Ashworth (1973), that in this organism differentiation occurs without feeding and so little confusion exists over cell and tissue growth. This has made it possible to isolate mutants of this organism in which the mutant gene is only expressed in the differential phase, thus providing further evidence for differential gene activity being implicated in the differentiation process.

If the process of differentiation is compared amongst the simpler eukaryotes, such as sponges, the higher fungi, coelenterates, and the larger algae, a number of important principles emerge. First, differentiation becomes entrenched as a process so that the organic form of the organism is determined by it. Secondly, as in the differentiation of *Dictyostelium*, the process becomes irreversible so that individual parts of the organism become entirely dependent on the presence and function of other parts. This reduction or loss of plasticity persists as a recurring feature throughout the organization of the more complex animals and plants, though some at least of these retain a capacity to repair

12

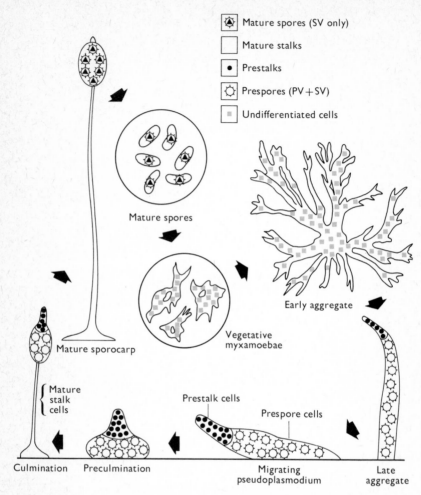

Mature spores (SV only)

Mature stalks

Prestalks

Prespores (PV + SV)

Undifferentiated cells

Mature spores

Early aggregate

Vegetative
myxamoebae

Mature sporocarp

{ Mature
  stalk
  cells

Prestalk cells

Prespore cells

Culmination   Preculmination

Migrating
pseudoplasmodium

Late
aggregate

Fig. 1.7 Stages in the life history and development of the slime mould *Dictyo-stelium discoideum*, showing the differentiation of the cells into spores and stalk. (After Gregg, J. H., 1971. *Developmental Biol.*, **26**, 479. Academic Press Inc.)

themselves or else to alter their form in the face of a changing en-vironmental demand.

A third feature of differentiation which is most apparent in the simpler eukaryotes is the stability which accompanies the differentiated state. This is nowhere better displayed than amongst the sponges (Porifera). As seen in Fig. 1.9, the organization of a sponge involves a number of different types of cells, each with rather different functions, i.e. skeletal, respiratory, feeding, and their arrangement in the whole sponge is related to their function. If an intact sponge is reduced to its single

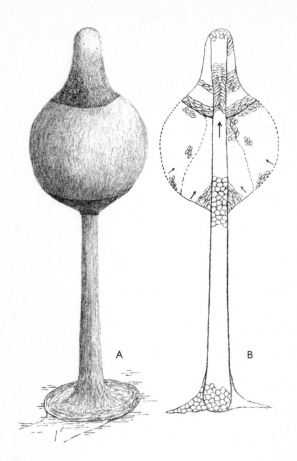

Fig. 1.8 Stalk formation in *Dictyostelium discoideum*. (A) Surface view of a late stage showing differentiation of spore and stalk areas. (B) Diagrammatic sectional view of A showing orientation of cells in stalk and fruiting head. The spore cells tend to differentiate from the periphery inwards, as indicated by arrows. (From Berrill, N. J., 1961. *Growth, Development and Pattern*. W. H. Freeman & Co.)

component cells by passage through a gauze sheet, the individual cells will reaggregate, first into small groups and finally into a complete sponge (Galtsoff, 1925). Present evidence suggests that the original cells retain their original functions and therefore that a sorting out process is involved.

The last important conclusion we should draw from studies of differentiation amongst simple eukaryotes is that, although relatively

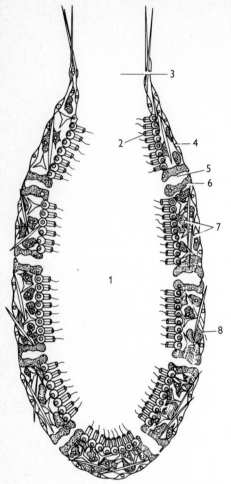

Fig. 1.9 Diagrammatic drawing of a simple sponge showing (1) cavity or spongocoel; (2) collar cells or choanocytes; (3) mouth or osculum; (4) epidermis; (5) pores; (6) pore cells or porocytes; (7) mesenchyme cells and spicules; and (8) skeletal elements. The flagella of the choanocytes create a current of water which enters via the pores and leaves via the osculum. (From Hyman, L., 1940. *The Invertebrates, Protozoa Through Ctenophora*, McGraw-Hill.)

undifferentiated stages of the life cycle may persist, the complex organism gradually becomes the dominant form. Amongst fungi, the simple mycelium remains an important part of the life of the fungus and the differentiated fruiting body is relatively transient. However, the simple multicellular green plants such as mosses, liverworts and ferns, and the simpler animal forms such as coelenterates, flatworms and rotifers, all exist as stable differentiated forms for the greater part of the life history. What may well have begun as a specialized mechanism for sporulation in slime moulds appears to have become the dominant differentiated phase in all higher organisms.

1.2 Stem cells and loss of plasticity

It should now be clear that one of the processes which accompanies differentiation is loss of cellular plasticity. The stem cells of the *Dictyostelium* fruiting body cannot regain the amoeboid state and they die after sporulation is completed (Bonner, 1967). Similarly, even simple eukaryotes, such as jellyfishes and seaweeds, have limited powers of regeneration. Although their loss of cellular plasticity is less marked than that found in more complex forms, the problems posed by their reduced powers of redifferentiation are quite evident. Complex organisms have attempted to counteract extreme loss of plasticity by retaining uncommitted stem cells which, when the occasion demands, can be directed along a required developmental pathway. This is seen clearly in the pluripotent interstitial cells of *Hydra* (Fig. 1.10) (Webster, 1971) and examples of the retention of this mechanism can be found in

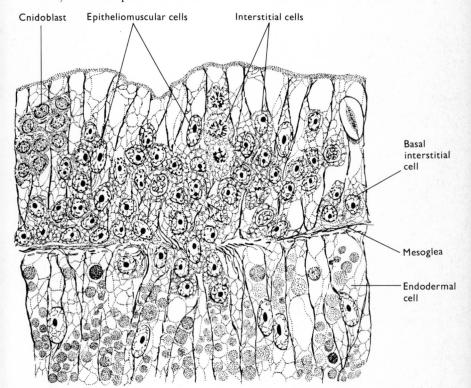

Cnidoblast   Epitheliomuscular cells   Interstitial cells

Basal interstitial cell

Mesoglea

Endodermal cell

Fig. 1.10   Onset of blastogenesis in *Hydra*, showing multiplication of basal ectodermal interstitial cells and invasion of epidermis (Blastogenesis is the development of a new individual by asexual reproduction from an existing cell mass). (After Brien, in Berrill, N. J., 1961. *Growth, Development and Pattern*, W. H. Freeman and Co.)

the relatively undifferentiated meristematic tissues of higher plants, in the retention of a separate germ cell line in many animals, and in the stem cells found in the erythroid tissues of vertebrates.

To take one example, the red blood cells of mammals constitute an essential part of their respiratory equipment and the animal dies in the absence of an adequate number of these cells. In an adult man approximately 2×10^{11} red cells are produced per day (Lajtha, 1964). The average red cell survival time is 110 days and the total red cell population in the human adult is about 25×10^{12}. Since the mature erythrocytes themselves do not divide and have, indeed, lost their nuclei, it follows that a population of precursor cells must exist, constantly releasing erythrocytes through differentiation. Moreover, if an animal becomes anaemic, the rate of red cell production increases until the anaemia is relieved. This constitutes a useful defence mechanism not only against anaemia-inducing diseases but against accidental blood loss through injury.

Characterization of the erythroid stem cell has not proved easy, particularly since one has to rely largely on morphology as a means of identifying one cell type from another. Small round cells, indistinguishable from the small blood lymphocytes, are common in adult bone marrow and display many of the characteristics of the elusive stem cell. The ability of such cells to form colonies of erythroid cells in the spleens of heavily irradiated mice, when administered by injection, gives support to this identification (McCulloch and Till, 1962; and De Gowin *et al.*, 1972). Stem cells do not directly produce erythrocytes; instead a graded series of cells can be observed (see Fig. 1.11) as follows: Stem cell → proerythroblast → basophilic erythroblast → polychromatophilic erythroblast → orthochromatic erythroblast → reticulocyte → erythrocyte. There is evidence that, with the development of the proerythroblast stage, the erythroid cell loses its capacity for self maintenance and is committed to die as a mature erythrocyte (Lajtha, 1964; Cole, 1975).

The simplest model which can account for the maintenance of the stem cell population, together with erythroid cell production, is that stem cells may embark on three distinct types of cell division each leading on to a different pathway. The three pathways determined by mitotic choice are: (1) a division of the stem cell into two daughter stem cells; (2) a division of the stem cell into one daughter stem cell and one proerythroblast; (3) a division of the stem cell into two proerythroblasts. Feedback signals involving the hormone erythropoietin probably help to determine the particular pathway chosen (Goldwasser, 1966). It is possible that the stem cell population is regulated by feed-back mechanisms sensitive to population size (Kivilaakso and Rytomaa, 1971), while demand for increased erythroid cell production would lead to rapid division of many stem cells along pathway (3).

It is interesting to notice that during ontogeny, the actual site of

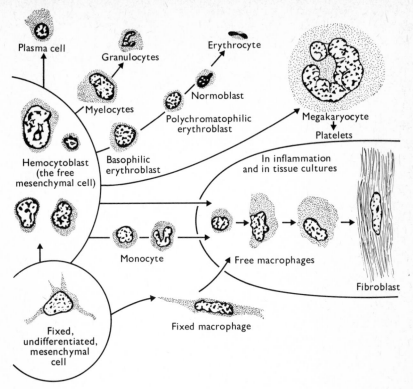

Fig. 1.11 Interrelationships of cells from human blood and some connective tissues. These cell types are similar in all mammals. The haemocytoblast cells include lymphocytes. (After Bloom, W. and Fawcett, D. W., 1969. *A Textbook of Histology*. W. B. Saunders & Co.)

erythropoiesis changes within any one species. The situation in the mouse has been intensively explored by Marks and his collaborators (Marks and Kovach, 1966). In the foetal mouse, the visceral yolk sac blood islands are the sites of red cell production until about day twelve of gestation, after which liver erythropoiesis, aided by the spleen, becomes dominant. Only on the sixteenth or seventeenth day of gestation does erythroid production become an activity of the bone marrow, but in the healthy adult the bone marrow is the sole site of erythropoiesis. However, if erythropoiesis is strongly stimulated in the adult mouse by hypoxia or erythropoietin administration, red cell production by the spleen is again detectable (Marks and Kovach, 1966). Presumably the spleen stem cells adhere to mitotic pathway (1) except in conditions of exceptional demand for erythroid cells. It should be stressed that this account of mammalian erythropoiesis is very oversimplified, and the possible differentiation of the stem cell into non erythroid cell lineages has been ignored. A more comprehensive account of mouse erythropoiesis and its control will be found in the paper of Cole (1975).

Although the existence and role of stem cells is perhaps clearer in mammalian erythropoiesis than anywhere else, it is likely that such stem cells are a common feature of the organization of animals and plants. Their presence has been postulated in systems as diverse as antibody-producing lymphoid cells and regenerating skin: the germ cell line can be viewed as essentially the basic stem cell line, retaining as it does a totipotency.

Stem cells can therefore be defined as relatively undifferentiated cells retained by the adult organism to offset some of the limitations imposed by tissue differentiation.

1.3 Cellular senescence and dedifferentiation

The loss of plasticity which cell differentiation often involves, has other implications for organisms besides those mentioned above. We have seen that the retention of a population of stem cells permits the continuous production of differentiated cells, these cells ultimately becoming redundant and expendable. But what of the tissues which do not retain stem cells? Are they incapable of continuous self replenishment? An adequate answer to this question is still only partially possible. There are three relevant points to consider.

First, many differentiated cells retain sufficient plasticity to permit mitosis and the production of similarly differentiated daughter cells. Explants of chick heart fibroblasts or mouse somite, for example, do retain their identity in tissue culture. Again, following satisfactory induction (Lash, 1968), the somite-derived cells persist in cartilage synthesis, a property never displayed by the fibroblasts. We should notice, however, that many cell strains probably survive in culture by cell selection, perhaps of 'stem cells', and indefinite growth in culture without cellular selection is still in question. Secondly, the loss of plasticity which accompanies differentiation is sometimes temporary or partial, and dedifferentiation is possible. Thus certain cells in the phloem of the carrot are capable of dedifferentiation with eventual production of a complete new carrot plant (Steward *et al.*, 1958), and many differentiated tissues lose much of their diagnostic characters when grown in tissue culture. Thirdly, and here we come to the crux of this section, some differentiated cells are apparently so specialized that dedifferentiation, and in some cases even mitosis, is impossible. Although in a simple multicellular organism like *Hydra* almost all the cells are capable of regenerating a complete organism (Webster, 1971), in the more complex eukaryotes such totipotency has been lost. Chick fibroblasts in culture produce further fibroblasts but not new chickens. Similarly, although some simple animals can regenerate whole limbs, amputated limbs are not regenerated by adult mammals. The most striking effect of extreme differentiation is seen in the neuron of the mammalian nervous system, a cell incapable of further division, so that throughout its life the organism is faced with a gradual and persistent loss of neurones. It is clear then that, in some tissues, if stem cells are not retained, the organism depends on the available differentiated cell population.

Now such a dependence on a 'closed' population of cells brings with it certain risks. It has even been suggested that non-disposable waste products may accumulate in the cells (Strehler, 1962), and deleterious somatic mutations may gradually impair the vital function of the particular tissue (Curtis and Crowley, 1963), though little direct evidence exists to support such a claim. What in some sense may be regarded as an opposite phenomenon may itself pose similar problems. Cell death is often employed as an important morphogenetic device by embryos and adults. Many secretory cells die in the process of releasing their products and the cells of the tadpole tail are destroyed and resorbed as part of the process of normal amphibian metamorphosis. Failure of certain of these cells to die may lead to developmental difficulties. For example, certain human individuals possess a genetic persistence of tissue between the toes. Such cells are programmed to die in the normal individual, and their survival leads to webbing of the feet.

Differentiation is therefore directly involved in the process of senescence, the ageing of the individual. Senescence should be viewed, not as a biological misfortune but as a highly programmed 'planned obsolescence' of the individual organism. Such a planned obsolescence ensures a rapidly turning-over population of individuals with a high level of genetic variation. Other processes are normally also utilized to accomplish senescence, but the reduced plasticity of the differentiated cell is certainly a key mechanism in the ageing phenomenon. Indeed, animals such as sea anemones which show little loss of cellular totipotency after differentiation, show almost no signs of an ageing process (Comfort, 1955).

1.4 Evolution of differentiation at the molecular level

Two cells are differentiated from one another if they have the same genome but the pattern of other molecules which they possess and synthesize is different. It follows from this statement that very simple cells may have a greatly reduced potential for differentiation. Indeed a population of cells might exist in which the genome was so simple that no cell could possibly survive without constantly expressing the full genetic complement. Such cells probably do not exist, but certainly within populations of bacteria the possibilities for differentiation are severely limited. Some bacteria may be induced by a substrate to transcribe a particular operon (See p. 70) while their sister cells, not exposed to the substrate, fail to express these genes. But such differences in metabolism are temporary and do not involve gross change of cellular morphology. We must also try to distinguish between changes which

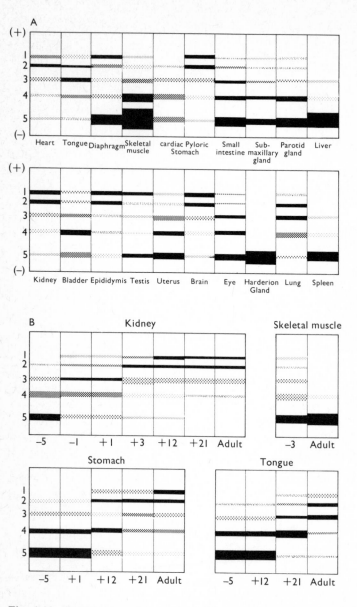

Fig. 1.12   Distribution of the enzyme lactate dehydrogenase (LDH) in the mouse. (A) LDH isozyme patterns of adult mouse tissues. (B) Ontogeny of LDH isozyme patterns in mouse tissues. See also Fig. 1.19. (After Market and Ursprung. 1962. *Dev. Biol.*, **5**, 368, Academic Press Inc.)

are simply part of the normal cell cycle, and changes which lead to more permanent differences between cells in the same population.

If a mammalian neuron and a cell from the intestinal epithelium are compared, striking morphological differences are at once obvious. These morphological differences are the visible evidence of more profound differences at the molecular level. No complete catalogue exists for all the molecular species within any one cell, but analysis of even the variant forms of one enzyme—lactate dehydrogenase—in different tissues of the mouse, reveals dramatic differences of distribution (Fig. 1.12). It is also obvious that the more biochemically complex an organism becomes at the genetic level, the greater are the possibilities for differing directions of cellular specialization. This is an important point to grasp, since it implies that complexity at the genetic level is a prerequisite for the multicellular complexity of organs and tissues. What is observed in the tissues of multicellular organisms is an interesting parallel with the results of evolutionary speciation. Just as selection has provided a range of organisms (species) which form interbreeding groups and are separated from similar groups by barriers to breeding, so differentiation produces not a vast array of different cells within one organism, but a limited number of rather homogenous cell populations or tissues. As with the individuals within a species, so with the constituent cells of a tissue, similarity but not absolute identity is the rule.

The simplest approach to studying the relationship between molecular evolution and differentiation, involves comparisons of the amino acid sequences in homologous proteins of different tissues and different organisms. Since the same tissue may vary in its form and function during development, this parameter must also be included in such a study. Such a detailed analysis has been undertaken with four principal proteins, viz., lactate dehydrogenase (LDH), haemoglobin, cytochrome-C and ribonuclease. Since molecules like haemoglobin and LDH have a quaternary structure involving the combination of different polypeptide monomers, the analysis has two aspects—firstly, what monomers are synthesized in a particular cell or tissue, and secondly, which combinations of monomers occur free in these cells and tissues? Such an approach is essential since different combinations of the same monomers may constitute an important aspect of tissue differentiation. What then has such an analysis shown?

(a) Cytochrome-C

Proteins such as cytochrome C, which play a crucial role in basic cellular metabolism, are present in every cell. Analysis of amino acid sequences of this molecule (Fig. 1.13), although revealing interesting evolutionary changes and relationships between different organisms (Fig. 1.14), and varying rates of molecular evolution (there is greater evolutionary divergence between human and tuna fish cytochrome C than between human and silkworm moth cytochrome C (Jukes and

22

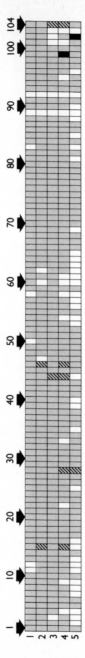

Fig. 1.13 Comparison of cytochrome C coding sequences in a variety of animals. (1) human; (2) horse; (3) chicken; (4) tuna fish, (5) yeast. White areas indicate amino acid differences from (1), cross hatched areas indicate differences from (1) which are common to two of the species, and dark areas indicate apparent gaps. (After Jukes, T. H., 1966. *Molecules and Evolution.* Columbia University Press.)

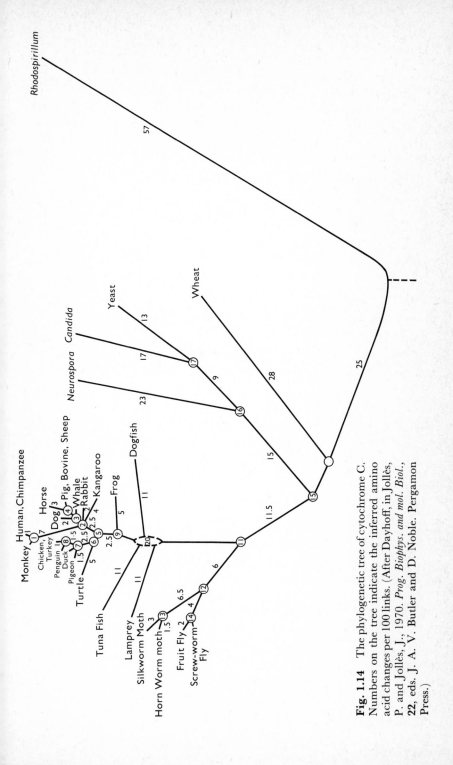

Fig. 1.14 The phylogenetic tree of cytochrome C. Numbers on the tree indicate the inferred amino acid changes per 100 links. (After Dayhoff, in Jollès, P. and Jollès, J., 1970. *Prog. Biophys. and mol. Biol.,* **22**, eds. J. A. V. Butler and D. Noble. Pergamon Press.)

24

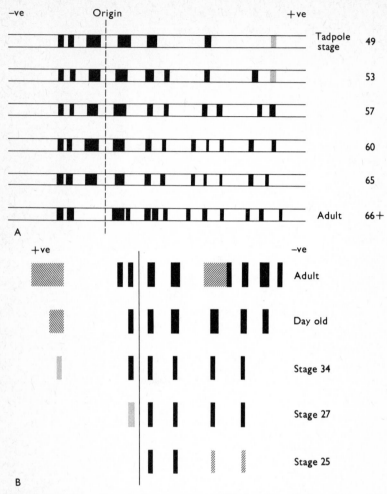

Fig. 1.15 Crystallins during ontogeny. (A) Starch gel electrophoresis of *Xenopus* crystallins. The major band present in tadpoles and absent in the adult is a γ crystallin. (B) Starch gel electrophoresis of embryonic chick crystallins. (After Clayton, M., 1970. *Curr. Top. Dev. Biol.*, **5**, 141. Academic Press Inc.)

Holmquist, 1972)) do not reveal tissue differences which might affect cellular differentiation. All the tissues of man and some closely related animal species harbour a single type of cytochrome C (Margoliash *et al.*, 1971). Only in such proteins as the crystallins of the lens of the eye (Fig. 1.15), the milk protein casein, or the proteins manufactured by the silk forming cells of the silkworm, is a highly unequal distribution revealed. These proteins are characteristic of particular tissues and are either scarce or absent in other tissues of the same organism.

This applies also to many other simple proteins on which evolutionary studies have been made. It is important, however, to draw conclusions from such data with caution. First, very few really critical comparisons have been undertaken of the same protein from differing tissues, and such as have been attempted are hampered by the shortcomings of the recognition assay used. Bands running in parallel on chromatographic and electrophoretic profiles do not necessarily indicate identical proteins. Secondly, the lower limits of detection are frequently far above that of one molecule per cell. When a tissue is said to lack a particular molecule, say casein or lens crystallin, this often more correctly means simply that none was detected—a very different observation!

(b) Haemoglobin

Structurally, the haemoglobin molecule of most vertebrates consists of a tetramer composed of two pairs of differing monomer molecules. Each monomer comprises a polypeptide globin chain of some 150 amino acids (Fig. 1.16), together with a haem group (Fig. 1.17). While the globins may vary from one haemoglobin to another, the haem moiety remains the same (see review by Maclean and Jurd, 1972).

As discussed earlier in this chapter (see p. 16) the sites of haemoglobin synthesis change during development, so that different tissues are involved in making haemoglobin at different periods of ontogeny. Moreover, the actual type of haemoglobin made at these developmental stages often varies, resulting in characteristic embryonic, foetal and adult haemoglobins. The human haemoglobin pattern is well understood and will serve as an adequate example. During early embryonic life multiple 'embryonic' haemoglobins (HbE) are present (Todd, Lai, Beaven and Huehns, 1970). These are composed of alpha, gamma, epsilon and zeta globin chains, such that their configuration, using the conventional system, may take one of the following five combinations —$alpha_2$ $epsilon_2$; $epsilon_4$; $alpha_2$ $zeta_2$; $epsilon_2$ $zeta_2$; and $zeta_2$ $gamma_2$. The embryonic haemoglobins are replaced by foetal haemoglobin (HbF) after about 3 months of embryonic life, and this molecule has the form $alpha_2$ $gamma_2$ as a globin arrangement (Itano, 1957). Shortly after birth, foetal haemoglobin is entirely replaced by adult haemoglobin, which occurs in two stable forms, HbA and HbA_2 with globin components $alpha_2$ $beta_2$ and $alpha_2$ $delta_2$ respectively. The proportion of the two adult haemoglobins is remarkably constant at 97.5% HbA and 2.5% HbA_2 in the healthy adult (Kunkel and Wallenius, 1955). Occasionally in conditions of anaemia or late pregnancy, some foetal haemoglobin reappears in the adult blood cells.

Since it is known that the different human globins are indeed the products of different genes and that their synthesis is under independent control (Allison, 1959), it follows that the erythroid tissues are expressing different globin genes at different times. Moreover, in conditions of stress (anaemia) normally silent genes may be re-expressed as, for

26

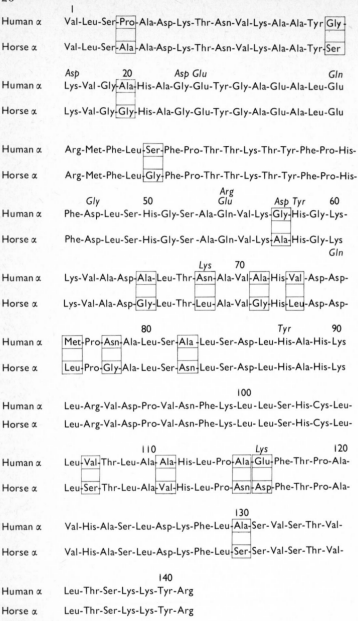

Fig. 1.16 Comparison of amino acid sequence of human and horse α globin chain. Known allelic substitutions of the human α chain are indicated in italics above the wild-type sequence, and one allelic and some non-allelic substitutions of the horse are indicated by italics below the wild-type sequence. (After Ohno, S. 1970. *Evolution by Gene Duplication*. George Allen & Unwin Ltd.)

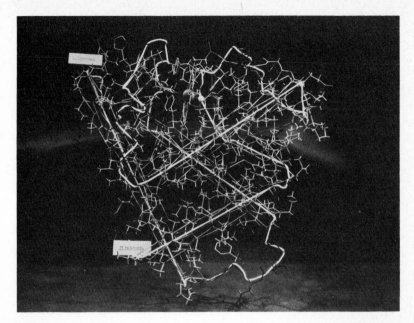

Fig. 1.17   Model of the myoglobin molecule, including side chains. The course of the main chain is indicated by a white cord and the iron atom is shown as a grey sphere. (From Kendrew, J.C. *et al.*, 1961. *Nature, Lond.*, **190**, 669.)

example, in the appearance of gamma globin, normally characteristic of the foetus, in adult humans. A similar phenomenon has been found in anaemic adult *Xenopus*, in which the tadpole haemoglobin reappears (Maclean and Jurd, 1971) (Fig. 1.18). While it is not at all obvious why differing globins are synthesized at different developmental stages or why multiple forms of globin are made at all (see p. 46 and discussion in Maclean and Jurd, 1972), it is clear that modulation of the globin genes is an example of an important differentiating system. Many cells in the vertebrate body make no globin within present limits of detection; others vary in the types of globin made, so demonstrating differentiation within a very restricted scale.

(c) Lactate dehydrogenase (LDH)

The work of Markert, Ursprung and their colleagues on LDH has provided an unrivalled knowledge of the distribution of one group of proteins (see discussion in Markert and Ursprung, 1971). LDH is an isoenzyme or isozyme; that is, it occurs in various tetrameric forms, rather like the variant forms of haemoglobin. Each monomeric subunit has a molecular weight of 35 000, giving a combined weight of 140 000 for the tetramer. Two common types of monomer occur, termed A and B, and these combine in various ways to yield the five common

28

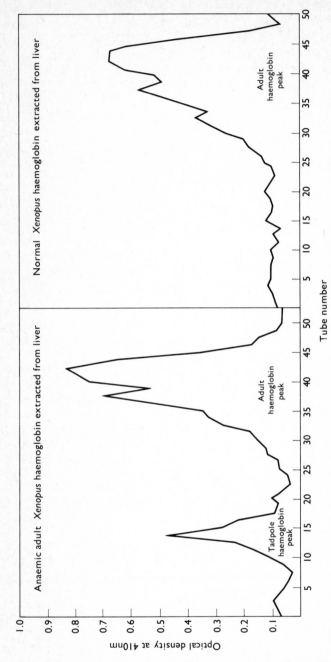

Fig. 1.18 Production of tadpole haemoglobin in an anaemic adult *Xenopus*. The haemoglobins have been eluted through carboxy methyl cellulose by a pH gradient.

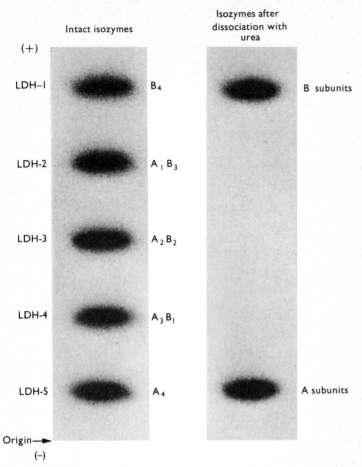

Fig. 1.19' Zymogram of LDH isozyme banding pattern on starch gel electrophoresis before and after dissociation in area. (After Market and Ursprung. 1962. *Dev. Biol.*, 5, 41, Academic Press Inc.)

isozymes of LDH (Fig. 1.19). The ratios of these five isozymes are highly characteristic for each type of tissue, and presumably the cytoplasmic milieu determines particular monomer combinations in particular tissues. Moreover, the rates of synthesis of the two subunits vary greatly in different tissues, indicating that genes are not just turned on or off, but that their expression is regulated by rate control of more than one hundredfold.

Three other significant conclusions emerge from Markert and Ursprung's experiments; these are illustrated in Fig. 1.12. First, different tissues in the same adult organism have widely different amounts of the five LDH isozymes. Secondly, during development, the same tissue has different tetramers present, apparently resulting largely from alterations in the rate of expression of the two subunit genes. Lastly, a third gene, making subunit C, is expressed in many vertebrates, but only in the cells of the primary spermatocytes. As with the globin genes, the LDH-C gene is quiescent in almost all the body tissues. The reasons for its highly restricted activity are not known.

Studies of LDH therefore provide very good information on the molecular basis of differentiation, and indicate both modulated gene activity and regulated cytoplasmic assembly of subunits as important mechanisms in the process.

TABLE 1.1   Nuclear DNA amounts in a variety of eukaryotes and prokaryotes. 1 picogram $= 10^{-12}$ g.

Organism	Amount of nuclear DNA		
Bacteriophage T_1	0.000007 picograms per particle		
,, T_4	0.000025	,,	,, ,,
E. coli	0.009	,,	,, cell
Bacillus megatherium	0.070	,,	,, ,,
Neurospora crassa	0.017	,,	,, ,,
Aspergillus nidulans	0.044	,,	,, ,,
Lupinus albus	2.3	,,	,, 2C nucleus
Rumex sanguineus	3.1	,,	,, ,,
Pisum sativum	9.1	,,	,, ,,
Zea mays	15.4	,,	,, ,,
Allium roseum	20.4	,,	,, ,,
Allium karataviense	45.4	,,	,, ,,
Lilium longiflorum	134	,,	,, ,,
Drosophila	0.2	,,	,, ,,
Chiton	1.3	,,	,, ,,
Nereid worm	2.9	,,	,, ,,
Squid	9.0	,,	,, ,,
Gryllus domesticus	12.0	,,	,, ,,
Protopterus	100.0	,,	,, ,,
Salmo irideus	4.9	,,	,, ,,
Esox lucius	1.7	,,	,, ,,
Tetrodon fluviatilis	1.0	,,	,, ,,
Amphiuma	168.0	,,	,, ,,
Rana esculenta	16.8	,,	,, ,,
Rana temporaria	10.9	,,	,, ,,
Alligator	5.0	,,	,, ,,
Black racer snake	2.9	,,	,, ,,
Pheasant	1.7	,,	,, ,,
Chicken	2.3	,,	,, ,,
Ox	6.4	,,	,, ,,
Man	6.0	,,	,, ,,
Mouse	5.0	,,	,, ,,

These observations on the synthesis and distribution of haemoglobin and LDH permit certain conclusions about the evolution of differentiation. These are that different genes coding for related forms of polypeptide chains provide opportunities for distinctions to arise between cells and tissues. Such distinctions sometimes involve a functional advantage, as in the metamorphic alteration in haemoglobin form with its intrinsic change in oxygen binding capacity in the bullfrog (Riggs, 1951). On other occasions they seem to be purely passive changes, for example, the presence of globin delta in haemoglobin A_2 in man. Since the role of genes making closely similar products offers such an important insight into the differentiation process, it is worth briefly discussing the significance of gene duplication in evolution.

1.5 Gene duplication in evolution and differentiation

If the haploid amounts of DNA for different organisms are compared (Table 1.1 and Fig. 1.20) a general trend of increasing DNA levels with increasing complexity is evident, provided one is prepared to overlook the extravagant amounts occurring in some lower vertebrate animals and liliaceous plants. Since the pattern of differentiation must presumably be dependent to some extent on such an increase, it is appropriate to ask how it was accomplished. The answer appears to be by

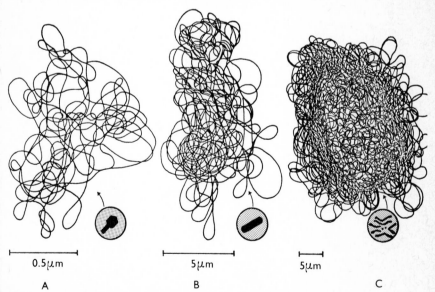

0.5 μm   5 μm   5 μm

A   B   C

Fig. 1.20 Diagrams redrawn from photographs showing the total length of DNA in (A) bacteriophage T_4, (B) the bacterium *E. coli*, and (C) the fruit fly *Drosophila*. (After Stahl, F. W., 1964. *The Mechanics of Inheritance*. Prentice Hall Inc.)

simple gene duplication (see discussion by Ohno, 1970; Watts, 1971). Unequal crossing-over during meiosis (Fig. 1.21) has been suggested as the mechanism responsible for the production of the different monomers in human haemoglobin (Lehmann and Carrell, 1969), but extra replication offers a simpler and much more direct means of gene duplication. I have suggested elsewhere (Maclean, 1973) a mechanism whereby accidental gene duplication could arise.

As originally suggested by Ingram (1961a), duplication of a basic globin gene was followed by a drift of the nucleotide sequences away from that of the original gene. The two genes might then become physically separated on the chromosome by inversion or other structural changes. The more recent the gene duplication, the closer together on the chromosome and the more similar in base sequence are the genes expected to be. Thus, similar globins have perhaps evolved from one another recently and their genes are actually more nearly adjacent on

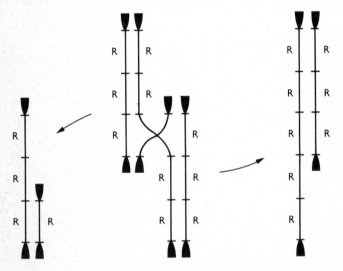

Fig. 1.21   Unequal crossing over between duplicated segments. R is, for example, the ribosomal transcription unit in the nucleolar organising region, and the figure shows only 3 of the numerous tandemly repeated copies. Middle column: First meiotic prophase. The homologous pairing between the duplicated segments in inexact. As a result, a chiasma is exchanged between the third gene of the chromosome at the left and the first gene of its homologue at the right. Left and right columns: Two daughter cells in 2nd meiosis. At the left, one of the two chromatids received the deleted nucleolar organizer (one R). At the right, one of the two received the further duplicated nucleolar organizer (five R's). As each crossing-over involves two of the four chromatids of the two homologues in pairing, two of the four gametes produced are affected (Ohno 1970). (After Ohno, S., 1970. *Evolution by Gene Duplication*. Springer Verlag and George Allen & Unwin Ltd.)

the chromosome than those of very dissimilar globins (Ingram, 1961a).

Whatever the mechanism involved, there seems little doubt that gene duplication is a major factor in the extension and evolution of the genome. It also ties in with our earlier discussion of LDH and haemoglobin. Some gene duplications might fortuitously provide deviant protein forms of real selective advantage, e.g. tadpole globin in the bullfrog (Riggs, 1951), but often the varied forms of the protein might be of no selective significance. In any event, as discussed for LDH, organisms seem very ready to exploit different forms of the same protein in different tissues, so leading to modified patterns of tissue differentiation.

2

Differential gene expression

Of no less importance than the identification of DNA as the genetic material is the realization that most cells of an organism have exactly the same genetic endowment. The evidence in support of this statement comes from two sources:

(a) As we have seen in Chapter 1, single differentiated cells from adult tissues can, occasionally, give rise to complete organisms (see p. 18).
(b) The DNA content of similar tissues from differing species vary widely, but provided cells in the same phase of the cell cycle are compared, the DNA content of individual cells of the same organism is often constant. Exceptions to this rule involve endopolyploid and polytene tissues on the one hand (see p. 63) and cells with differential amplification on the other (see p. 95).

To the extent that the rule of the equivalence of the genetic endowment of different cells in the same organism does obtain, it becomes clear that we must account for the differences between differentiated tissues in terms of *the varying expression of the same pool of genetic information*. There is no doubt that cells within the same organism, which have become differentiated from one another, do express different populations of genes in their protein content and synthesis. What is less obvious is how this state of affairs comes about.

In this chapter, we will outline the processes by which cells control the activity of their own genetic material and its expression in RNA and protein synthesis. We will do so by elaborating answers to the following key questions which are involved in this phenomenon, namely:

1 How do cellular differences arise during normal embryonic development, or in the growth of organisms from single cells?
2 Following early development, are all adult tissues committed cell populations, only ever expressing some genes and not others in their cells? When new proteins appear in response to changes in the environment, are they due to development of new lines of cells from stem cells?

3 Given that populations of differentiated cells can alter their expression, does this necessarily involve cell division?

4 If changes in gene expression occur between cell divisions, how are they controlled? Such changes are normally viewed at the level of the protein product but are they transcriptional or post-transcriptional?

5 What are the mechanisms for controlling gene expression?

The first four of these questions will be discussed in this chapter leaving the fifth to be dealt with in Chapter 3.

2.1 How do cellular differences arise during normal embryonic development, or in the growth of organisms from single cells?

A study of the growth of an organism from a single fertilized egg to give a differentiated series of cell populations has occupied embryologists for about a century and it is now possible to take a long view of many years of careful investigation and attempt some synthesis of ideas. The greatest difficulty has been in understanding the relationship between what have come to be called *mosaic* and *regulative* development.

The earliest experiments, carried out by the German zoologist August Weismann and others, suggested that the differing groups of cells in an embryo owed their origin to different parts of the original fertilized egg. As the egg divided by symmetrical cleavage along well defined planes, so the different parts of the egg cytoplasm became isolated from one another. It seemed legitimate to conclude that the mosaic of the eventual embryo, comprising recognizably different cell types in its diverse tissues, had arisen from an inherently mosaic egg (Fig. 2.1). Although the egg might look rather uniform, it was, in fact, already subdivided into differentiated areas. Early experiments supported this view. The theory of the mosaic egg would predict that the four blastomere cells resulting from two rounds of cell cleavage were actually different in their properties and potential, although in many embryos they were outwardly identical. Death or destruction of one blastomere would thus lead to the development of a defective embryo. And so it proved to be. In 1888, an important experiment by the German anatomist, Roux, involved killing one of the two blastomeres of a two-cell stage frog egg with a hot needle. The remaining cell developed into only half a tadpole. A related experiment of Spemann's in 1938 is illustrated in Fig. 2.2.

However, only a few years later, the experiments of Driesch pointed to a different conclusion. Working with the sea urchin, Driesch separated the cells of cleaved eggs and found that single cells derived from even 4-cell stage blastulae would grow on to produce normal larvae. So for

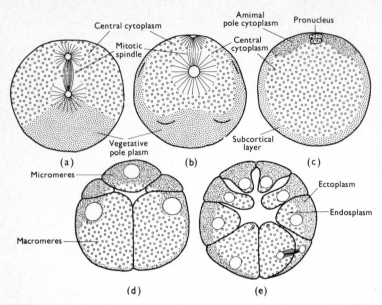

Fig. 2.1 Diagrams showing the segregation of the egg cytoplasm in early development of the snail *Limnaea stagnalis* (a) egg immediately after laying, (b) extension of vegetal pole cytoplasm, (c) formation of animal pole cytoplasm, (d) eight cell stage, showing micromeres, and (e) the blastula, with differential distribution of animal and vegetal pole cytoplasm. (After Raven, C.P. *Differentiation in Mollusc Eggs*, In Cell Differentiation, *Symp. Soc. exp. Biol.*, **17**, 275. Cambridge University Press.)

seventy years the pendulum has swung back and forth between interpreting all eggs as mosaic or no eggs as truly mosaic. Added to this, the literature has been complicated by the confused use of the opposing terms *mosaic* and *regulative*, the one normally referring to predetermined differentiation and the other to the retention of totipotency, that is the ability to go in any one of a number of different developmental directions. Unfortunately these terms have come to be used of embryos as well as eggs and it is not always clear in some texts whether a mosaic embryo is one arising from a mosaic egg or one in which all or almost all regulative ability has been lost. This question is thoroughly discussed in an excellent book by Davidson (Gene activity in early development, 1968), where the point is well made that all eggs are partly mosaic and partly regulative.

(a) Mosaic and regulative development

We must now face up to the question of how exactly regulative and mosaic development works. What factors influence the different cells of a maturing highly regulative embryo in bringing about the final

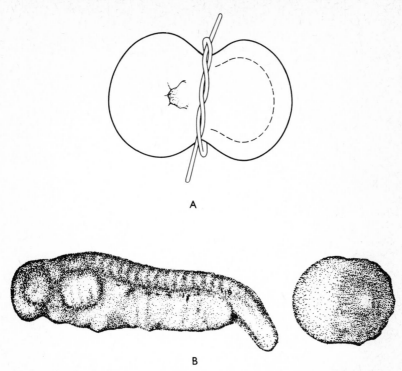

Fig. 2.2 Effects of a ligature applied to a two-cell stage newt embryo. (A) Ligature in the early gastrula, having been applied just behind the first cleavage furrow. (B) Twins resulting from A above. The dorsal half which included the first furrow develops into a well-proportioned embryo, the ventral half forms only a 'belly piece'. (After Tiedmann, H., 1967. 'Primary induction and determination' In *The Biochemistry of Animal Development*. ed. R. Weber, *Biochemical Control Mechanisms and Adaptations in Development*, Vol II. Academic Press Inc.)

development of a normal organism? And conversely, given that certain eggs are apparently highly mosaic, even before the first cleavage, what is the nature of the differences between the different cytoplasmic sections which, after the partition of cleavage, ensure that one blastomere will develop in one direction and a sister blastomere in another quite different direction?

(b) Regulative development

Embryos arising from strongly regulative eggs retain the plasticity (or totipotency) of their individual cells to a late stage. Thus the blastomeres of many animals, such as the sea urchin *Echinus* or the Cephalochordate *Amphioxus*, will, if dissected away from the other blastomeres, proceed to develop into complete normal organisms. An excellent assay for regulative ability in older embryos is provided by transplantation experiments. If a *fate map* is constructed for an embryo (Fig. 2.3), in which the embryonic cells are named according to the tissue into which they would normally develop, we can explore the possibilities of removing a prospective 'eye' cell, i.e. an embryonic cell which, if left *in situ* would, along with its neighbouring cells, develop into an eye, and transplanting it into an area of prospective muscle cells in an embryo. Will the transplated cell then develop into eye or muscle tissue? Many early embryologists, such as Vogt and Spemann, carried out such experiments. Their findings were that in embryos from highly regulative eggs, provided the cells were transplanted at an early stage, they would regulate to the fate of the site into which they were transplanted. But, if the experiment were repeated with older embryos, the cell failed to regulate and persisted in demonstrating the fate to which it had been originally committed. The significant event in fixing cell fate appeared to be associated with the development of a part of the embryo known as the dorsal lip of the blastopore.

The blastopore is a hole which forms in the hollow sphere of cells which comprises the blastula, and through which the cells of the blastula proceed to migrate to the inside of the hollow ball, thus forming an invaginated cellular lining enclosing the archenteron. The embryo at this stage is termed a gastrula. It appears that, as the cells on the surface of the blastula roll in over and through the dorsal lip of the blastopore, they become committed to their 'fate' and lose their regulative capacity or totipotency. This inductive capacity is retained even if the dorsal lip is transplanted (see Fig. 2.4). Only prior to gastrulation can the cells of embryos derived from regulative eggs be induced to accept a new

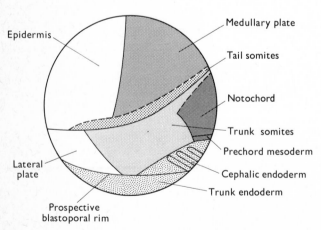

Fig. 2.3 Fate map of the early gastrula of the axolotl. (After Holtfietet, J. and Hamburger, V., 1955. In *Analysis of Development*. W. B. Saunders and Co.)

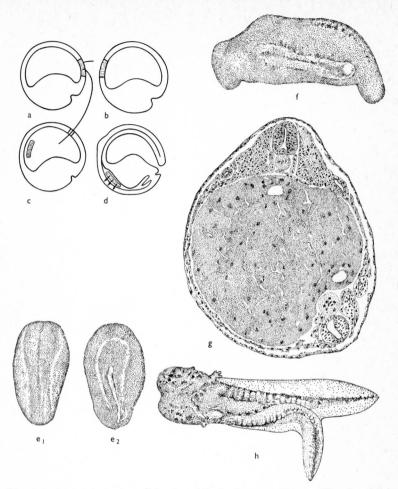

Fig. 2.4 Transplantation of the dorsal lip in the newt *Triturus*. The dorsal lip can be removed from (a) and implanted into ventral ectoderm (b), into the blastocoel (c), or next to the ectoderm (d). A normal neurula is shown in (e_1) and another neurula (e_2) which, after dorsal lip implantation, has a secondary neural plate. In (f) is seen an embryo carrying a second developing embryo on its side, complete with somites and otic vesicle, and a cross section of such an embryo is shown in (g). An embryo with a very well developed secondary twin is shown in (h), again resulting from dorsal lip transplantation. (After Kühn, A., 1971. *Lectures on Developmental Physiology*, 2nd edn. Springer Verlag.)

fate by transplantation. The discovery of the controlling role of the blastopore dorsal lip led to an eager search for its mode of action and, in particular, for a substance which it might release which would have the function of finalizing cell commitment. Such a substance was, in the then current hypotheses, termed the inductor. We will return to a discussion of the inductor after a look at development in strongly mosaic eggs.

(c) Mosaic development

The characteristics of the strongly mosaic egg are well illustrated by the ascidian (sea squirt) *Styela*. The egg is complexly pigmented and, as a single cell, shows a characteristic distribution of the various pigments. In a classical study of the development of this embryo, published in 1905, Conklin followed the placement of these pigments in the developing gastrula, and showed that particular areas of the cytoplasm of the egg were destined to develop into particular embryological features (see Fig. 2.5). So, in a variety of organisms, of which the snail *Ilyanassa*, the nematode *Ascaris*, and the annelid *Nereis* are additional examples, development is strongly mosaic and if single blastomeres are separated from their sister cells, they will proceed to mature by cell division and growth into only that part of the embryo which a fate map suggests as their normal resultant tissue.

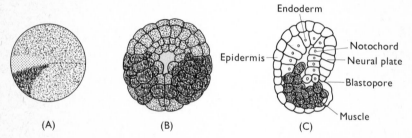

Fig. 2.5   Localization of the yellow pigment in *Styela*. (A) lateral view of fertilized egg, (B) posterior view of early gastrula (C) longitudinal section of late gastrula. The pigmented area is darkly coloured. (After Waddington, C. H. 1956. *Principle of Embryology*. George Allen & Unwin Ltd.)

To return now to our question, what unifying explanation can we find for the following observations?

1 Strongly mosaic eggs apparently predetermine the fate of a cell by providing it with predifferentiated egg cytoplasm (see also p. 117).
2 Strongly regulative eggs yield embryos which lose their regulative ability after gastrulation and are then committed to a fate just as final as that of cells from mosaic eggs.
3 Cells isolated from pregastrulation embryos from regulative eggs can either develop into whole organisms if reared alone or, adopt the 'tissue-style' of the new site if transplanted to another part of the embryo.

The logical conclusion is that all eggs have highly partitioned cytoplasm and all blastomeres, although apparently alike, are different. In a word, all eggs are to a considerable extent mosaic. But blastomere cells from so-called regulative eggs are not yet fixed to their fate, and in the event of isolation or transplantation can adjust to the new circumstances imposed by the lack of sister cells or the presence of new surrounding cells. The function of the regulator, the dorsal lip being the primary one, must then be seen as that of throwing a pre-set switch and finalizing a potential commitment. We will not then be surprised to find that the search for the nature of the inductor has proved to be extremely difficult. There is still no clear-cut conclusion regarding the nature of the specific inductor which is released from the dorsal lip of the blastopore since fractions isolated from a wide variety of biological sources prove to have inductive capacity. No doubt the embryonic cells are poised delicately, awaiting a relatively simple signal which initiates the realization of their commitment. And, of course, once a heterogeneous cell population is established, there is little difficulty in maintaining and extending differentiation. The controlled availability of metabolites or hormones are obvious ways of following up early differentiation and gradually deepening the commitment of individual cell populations to their particular destinies.

Leaving aside for the moment interesting questions relating to the nature of the initial differentiation of the egg cytoplasm, which will be discussed more fully in Chapter 4, we will anticipate some of the conclusions to be reached in that chapter and assume that the embryonic cell cytoplasm does have a profound influence over the daughter cells and the genes contained in them. While appreciating, then, that the cytoplasm has an important role both in initial commitment and in later regulation of the genetic material, we can proceed to examine genetic regulation in the committed cell, both embryonic and adult.

(d) Growth of organisms from single cells

Outside the field of embryology it is not easy to find equivalent examples of single cells giving rise to whole organisms or at least to complex differentiated cell systems. Most cells apparently lose their effective, if not their potential, totipotency during growth and development. Notable exceptions are found amongst certain cells in the phloem of the carrot (Steward, 1968) or in callus tissue of plants such as tobacco *Nicotiana* (Fig. 2.6).

Apart from the remarkable dedifferentiation which occurs in such experiments and their implications for the reversibility of the differentiated state (See also p. 188), these examples highlight again how a mosaic of different cells may sometimes develop from one primary cell. Unfortunately, neither system has been widely studied in these terms.

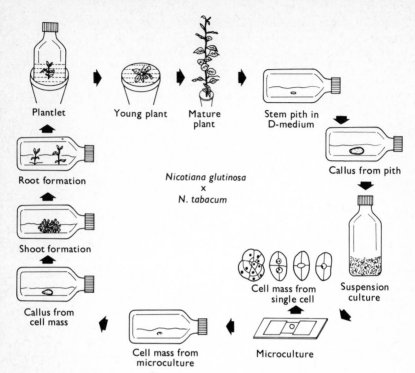

Fig. 2.6 Diagram of development of normal tobacco hybrid plants from individual cells isolated from fresh stem callus. (After Vasil V. and Hildebrandt, A. C. 1967. *Further Studies on the Growth and Differentiation of Single Isolated Cells of Tobacco in vitro.* Vol. 75. 139–51. Springer Verlag.)

2.2 Following early development, are all adult tissues committed cell populations only ever expressing some genes and not others in their cells? When new proteins appear in response to changes in the environment are they due to development of new lines of cells from stem cells?

The first part of our second question must be answered at two levels. To a very great extent, adult tissues *are* committed cell populations, expressing only a confined part of their total genetic endowment. Muscle cells are, and continue to be muscle cells, and liver cells continue to function as liver cells, throughout the life of the organism. If a search is made for lens crystallins in liver cells or muscle cells, rather than in the eye, they will almost certainly be found lacking. Similarly, it is vain to look for evidence of ciliation in neurones or of expression of genes for globin in epithelial cells. We will return to the question of whether these proteins are absolutely lacking from these cells in answering question 5. Some gene products are unquestionably widespread if not universal, so that all cells are active in expressing genes for transfer RNA or ribonuclease, whether they are liver cells or neurones, but other genes are expressed only in restricted tissues, and these tissues and their cells do apparently always express these genes.

This state of apparent fixation of cells to a particular pattern of genetic expression is verified by the formation of tissue cultures from explanted cells. A wide variety of differentiated cell types retain their tissue specificity in terms of cellular morphology and the spectrum of proteins synthesized even after prolonged growth and cell division in tissue culture. This stability of the particular spectrum of expression is well illustrated by the imaginal discs of insects and which have been particularly well studied in the fruit fly *Drosophila* (Hadorn, 1967). Imaginal discs are groups of cells distributed in different parts of the larva (See Fig. 2.7) which, if removed before their overt differentiation in the larva, can be propagated in an apparently undifferentiated state for several generations in the haemocoels of adult insects (Fig. 2.8) but, when returned to a larva which is permitted to metamorphose, the disc

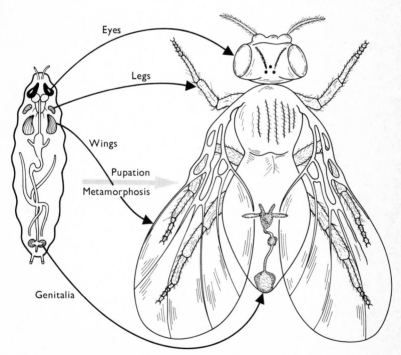

Fig. 2.7 Development of *Drosophila* imaginal discs, their locations in the larva and the adult tissues into which they develop. (After Markert, C. L. and Ursprung, H. 1971. *Developmental Genetics*. Prentice Hall Inc.)

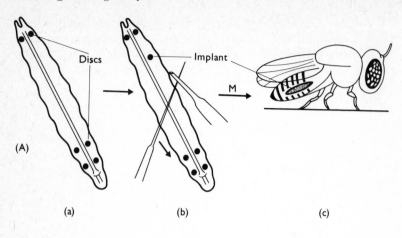

(A)

(a)   (b)   (c)

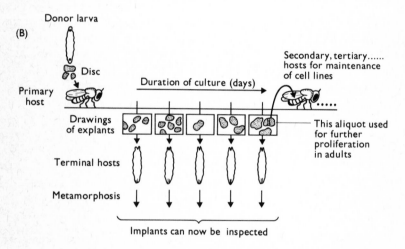

Fig. 2.8 (A) Transplantation of imaginal discs. Imaginal discs are dissected free from a donor larva (a) and pieces implanted in another larva (b) or adult (c); in both situations the pieces will grow but not differentiate. (B) Proliferation of imaginal discs in host files. Disc explants may be inserted into larval host which, following metamorphis, will reveal the expression of the disc in the adult. (After Markert, C.L., and Ursprung, H. 1971. *Developmental Genetics*. Prentice Hall Inc.)

cells differentiate into the precise tissue of their original determination (Fig. 2.9). Imaginal discs demonstrate very forcibly the long lag which may impose between determination (the state of commitment to a differentiated fate) and the appearance of the overt differentiation. The cells of imaginal discs in the larva are embryological in character and in no way betray the ultimate differentiated form which is their fate.

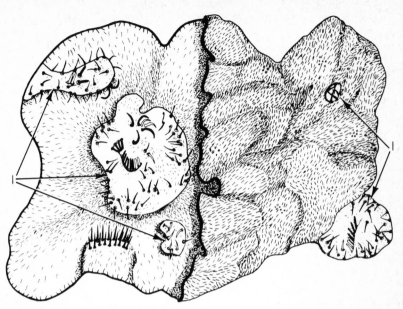

Fig. 2.9 Implanted imaginal discs in *Drosophila*. The figure shows a diagrammatic representation of an implant containing both wing and leg elements. The implant has been opened on the left side, exposing bristles and hairs on the inside surface of the vesicles. Several leg areas are shown (l) some with bristles of two phenotypes and surrounded by wing tissue. (After Poody, C.A. *et al.* 1971. *Dev. Biol.*, **26**, 467. Academic Press Inc.)

This point will be discussed further in chapter 9, together with the rare but significant phenomenon of transdetermination which has been observed in this system. But, to return to our present point, in the case of imaginal discs, the repeated cell division in no way dilutes restrictions of gene expression which were imposed on the cells early in development. Or, to put it in another way, embryological determination may often precede the overt differentiation, in this case largely under hormonal control. On the other hand within the confines of muscle remaining muscle, or liver cells remaining liver cells, some latitude of gene expression can occur. During starvation the population of liver enzymes in the rat is markedly different from that of the normal unstarved liver (Table 2.1) and the form and activity of mammary gland tissue varies greatly under the influence of the hormones controlling the female sexual cycles.

An excellent example of the latitude permitted in normal differentiated adult tissue involves the haemoglobins present in human adult erythrocytes. There are two main forms of human haemoglobin and a number of minor components. As we have seen earlier (See p. 25) the two main human haemoglobins are normally called adult and foetal,

Table 2.1   Effect of 7-day fasting on liver components of the rat* (Allard *et al.*, 1957)

Components	Fed controls		Fasted		Per cent loss per nucleus
	Wet weight	Units/nucleus	Wet weight	Units/nucleus	
Initial animal weight, g	230 ±2		228 ±2		
Final animal weight, g	264 ±5		148 ±9		
Final liver weight, g	8.9 ±0.4		3.9 ±0.7		
Final liver/ body weight ratio × 100	3.37 ±0.25		2.64 ±0.38		
Cellularity: nuclei/ g × 10^{-6}	204 ±40		394 ±74		
Mitochondriality: Mitochondria/ g × 10^{-10}	31.1 ±1.6	1525	32.2 ±2.9	817	46
RNA, mg/g	7.3	35.8 × 10^{-9}	6.8	17.3 × 10^{-9}	52
Nitrogen, mg/g	31.9 ±0.5	156.0 × 10^{-9}	38.8 ±1.4	98 × 10^{-9}	37
GDHase	0.291 ±0.032	14.4 × 10^{-9}	0.495 ±0.076	13.8 × 10^{-9}	4
Guanase	0.083 ±0.006	4.12 × 10^{-9}	0.127 ±0.007	3.31 × 10^{-9}	20
Alkaline RNAase	680 ±202	3.33 × 10^{-6}	1109 ±25	2.81 × 10^{-6}	16
Acid RNAase	338 ±68	1.66 × 10^{-6}	504 ±91	1.28 × 10^{-6}	23
Acid Pase	1122 ±157	55.0 × 10^{-6}	1624 ±250	41.2 × 10^{-6}	25
Alkaline Pase	164 ±60	8.0 × 10^{-6}	213 ±52	5.4 × 10^{-6}	32
Uricase	0.111 ±0.019	5.71 × 10^{-9}	0.110 ±0.008	2.89 × 10^{-9}	49
Inosine Phosphorylase	1.96 ±0.15	99.6 × 10^{-9}	1.80 ±0.13	50.0 × 10^{-9}	50
ATPase	1472 ±120	72.1 × 10^{-6}	1418 ±108	36.9 × 10^{-6}	49
Cathepsin	0.096 ±0.040	4.7 × 10^{-9}	0.089 ±0.018	2.26 × 10^{-9}	52
Xanthine oxidase	0.018 ±0.005	0.86 × 10^{-9}	0.004 ±0.001	0.100 × 10^{-9}	88

* Mean and S.D. of four pools of four livers.

each consisting of four globin chains joined to give a large tetrameric molecule. In each haemoglobin type, the globins are of two kinds, foetal haemoglobin consisting of two alpha chains and two gamma chains, and adult haemoglobin of two alpha chains and two beta chains. Alpha, beta and gamma globins are known to be coded by different genes (Ingram, 1961b). As a baby matures into childhood, the foetal haemoglobin, the main constituent of red blood cells before birth, gradually disappears, to be replaced by adult haemoglobin, and the normal adult has no detectable foetal haemoglobin in the circulating red cells, i.e. the gene for gamma globin is not expressed in the adult erythrocyte. However, in some cases of acquired anaemia, and not infrequently during pregnancy, foetal haemoglobin reappears in the circulating cells. Moreover, as we shall see shortly, there is good evidence that the gamma gene is expressed, along with the beta gene, in a large number of the red cells. From this we can conclude that, although human erythrocytes do not appear to synthesize such alien substances as, say, insulin, or their precursors, they can, in certain circumstances, vary their genetic expression within narrow limits.

The evidence for the wide dispersal of the foetal haemoglobin amongst the erythrocytes of the anaemic or the pregnant mammal contradicts any assumption that a new population of cells, derived from a small clone of cells which had somehow retained the potential for making the foetal (and not the adult) form, was present in the blood of these individuals. There are two pertinent points. First, by the application of a staining technique dependent on the differential extraction of the different haemoglobins by strong acid or alkali, it appears that, in the blood of anaemic adults (or the blood of new-born babies, in which the haemoglobin transition is underway), many red cells are intermediate in their staining properties between the foetal and adult appearance, i.e. they have a mixture of both haemoglobins (O'Brien, 1961). The second line of evidence is drawn from individuals suffering from sickle cell anaemia, a genetic disorder caused by an altered amino acid in the beta globin chain. Affected red blood cells become sickle shaped when exposed to reduced oxygen tensions (Fig. 2.10). Since only the adult form of human haemoglobin possesses the beta globin chain, only cells containing the adult haemoglobin should sickle. For the present case, a fortunate side effect of the disease is a retained high level of foetal haemoglobin. We can now ask what happens when the blood of small children, suffering from sickle cell anaemia, is exposed to lowered oxygen levels? The answer is that all of the cells can be induced to take up a sickled shape (Schneider and Haggard, 1955). It can safely be concluded then, that all the cells contain beta globin, and it is known from examination of the extracted haemoglobin, that many of the cells

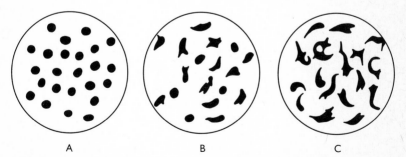

A   B   C

Fig. 2.10 Silhouettes of human red blood cells, (A) normal in a normal homozygote, (B) sickle-cell trait in a mutant heterozygote and (C) sickle-cell anaemia in mutant homozygote. (After Herskowitz, I. H. 1973. *Principles of Genetics.* Collier-Macmillan Ltd.)

must also contain gamma globin. It appears, therefore, that although red blood cells normally express either the beta or the gamma globin gene, the simultaneous expression of both is possible.

There is other evidence for apparent gene switching in differentiated tissues from studies on the puffing phenomenon in the giant chromosomes of dipterous flies, and also for changes in the levels of certain enzymes during the cell cycle of yeast cells, but we can more profitably consider these cases in relation to our next question. Of course the dramatic example of gene induction in bacteria must not be overlooked, in which, prior to induction of the enzyme β-galactosidase by its inducer (see p. 70), less than one molecule of the β-galactosidase messenger RNA is synthesized per cell per generation (Beckwith and Zipser, 1970). To date, the best evidence for a dramatic transcriptional response to a specific inducer is probably the synthesis of ovalbumin by chick oviduct in response to added oestrogen. Cox *et al.* (1974) have shown that the oviduct cells prior to induction have only one or two molecules of ovalbumin messenger RNA, and this rises dramatically to 36,000 after induction!

Having established the point that some latitude of genetic expression is permitted to differentiated adult tissues, it is important to emphasize that in some cell types almost no latitude is allowed, and synthesis of a new and different protein by the tissue demands the growth of a new line of cells. No better example of this confined type of genetic expression can be found than in the synthesis of immunoglobulins.

Probably all vertebrate animals, and some invertebrates, possess lymphoid cells responsible for the production of gamma globulin. This protein is active as an antibody in response to a highly specific target molecule, most frequently of foreign origin, termed the antigen. It now appears likely that a given lymphoid cell can synthesize only a few types of immunoglobulin, perhaps only one. In the event of the animal body being invaded by a new antigen, not previously encountered, new specific antibody must be assembled. The rather slow appearance of this antibody on the initial encounter with the antigen certainly suggests that existing cells cannot switch directly to its production. On repeated challenge of the animal with doses of the same antigen, more and more antibody is assembled, suggesting the development of a specific antibody-producing cell line. There is now very strong evidence for the development of such clones of cells in the spleen and other lymphoid organs of animals challenged with novel antigen, and the initial low antibody activity certainly points to their growth from very few primary stem cells.

The committed or uncommitted nature of these stem cells has become a matter of great concern to immunologists, leading to the elegant 'clonal selection' theory of antibody formation advanced by Sir Macfarlane Burnet in 1959. It is sufficient to notice here that, in the lymphoid tissues, demand for a new immunoglobulin is not answered by altering the genetic expression of existing cells and tissues, but by the comparatively slow growth from a few primary cells of a new cell population dedicated to the production of the protein needed.

Indeed, some particularly careful work by Nossal and Lederberg (1958) has shown that probably each cell synthesizing gamma globulin makes only one species of antibody. These workers were able to isolate single cells from the lymph nodes of rats previously exposed to repeated doses of flagella protein from strains of the bacterium *Salmonella*. If the motile bacteria encounter antibody made against their flagella, they are immobilized. By establishing isolated lymphoid cells in hanging drop cultures, Nossal and Lederberg were able to expose different strains of *Salmonella* to the drop of culture fluid surrounding a given cell: if the bacteria were immobilized, the cells were known to have released antibody specific to its flagellar surface. From these experiments there was no evidence that any individual lymphoid cell was programmed to make more than one antibody type. A similar experiment is illustrated in Fig. 2.11.

While discussing the response of the vertebrate body to a challenge with foreign antigen, it is well to notice that this system provides an excellent example of permanent alteration in the character of an organism in response to an often trivial environmental factor. Even monozygotic twins must be expected to become biologically and permanently distinct as a result of the differing antigens which they accidentally encounter whether in the same or in different environments. As is argued so well in the essays of Sir Peter Medawar (1957), the individual organism, if not always genetically unique, rapidly becomes phenotypically unique as a result of the permanent changes which accrue from fortuitous immunological encounters.

The answer to our question seems to be that, normally, all cells have a restricted capacity for altering their genetic expression, and there is an apparent long-term control which keeps muscle tissue as muscle even after prolonged growth in tissue culture. But, within these confines, *some* cells have *some* freedom to alter the pattern of proteins synthesized. Whether this is operated by switching genes on and off will be discussed in answering question 5 (see p. 63).

2.3 Given that populations of differentiated cells can alter their expression, does this necessarily involve cell division?

Although this question is often asked in these terms, in fact it subdivides into two separate questions. The first asks whether alteration in gene expression requires DNA synthesis, and the second whether it demands DNA synthesis plus cell division. All cells pass through a sequence of stages which, *in toto*, are referred to as the cell cycle. Immediately after mitosis daughter cells are said to be in G_1 (Fig. 2.12). Such cells are said to have a 2C-DNA value which is twice the amount

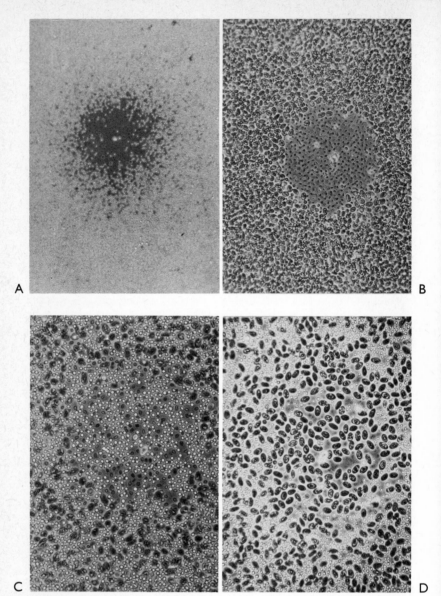

Fig. 2.11 The assay of antibodies released by individual spleen cells from mice immunized with a mixture of sheep and chicken red blood cells. Typical plaques produced by incubating mouse spleen cells with (A) sheep red blood cells (SRBC) only; (B) chick red blood cells (CRBC) only; (C) and (D) a mixture of SRBC and CRBC. In (C) only SRBC have lysed, leaving an area dominated by intact CRBC. In (D) only CRBC have lysed, leaving an area dominated by CRBC nuclei in ghost cells and unlysed SRBC. Note that each cell will lyse only one red cell type, presumably because it manufactures only one species of antibody. (Photographs kindly provided by Dr A. E. Wild.)

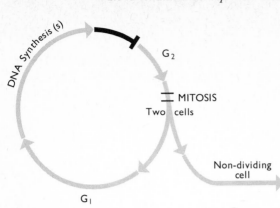

Fig. 2.12   Diagram of the cell cycle (for explanation see text). (After Baserga, R. 1965. *Cancer Res.*, **25**, 583. Cancer Research Inc.)

present in the gametes (1C). Some differentiated cells remain permanently in G_1, for example, the nucleated erythrocytes of frogs and birds and the neurones of all tetrapods. However, in tissues which do divide, G_1 is followed by an S period, during which DNA synthesis occurs, and after completion of S, a G_2 period may intervene prior to mitosis and cell division (Fig. 2.13). Cells in G_2 have a 4C-DNA value. To rephrase the questions then, can cells in G_1 alter their genetic expression prior to going through an S phase, and secondly is S, G_2, and a return to a new G_1 through cell division, necessary for a change in expression?

A complication arises if we are studying only changes in enzyme levels. With the available detection techniques it is not always easy to distinguish between increased synthesis of an enzyme, and its sudden appearance *de novo*. Cells in G_2 have twice the DNA content of those in G_1 and in diploid systems G_1 cells already possesses two copies of each gene locus since each chromosome is represented twice. If the enzyme level is critically affected by the number of gene copies, we might expect a rather sudden increase in that level after S. Moreover, if the S period is attentuated, increases in the levels of different enzymes may be characteristic of a large part of the entire cell cycle.

Because specific gene expression can often be monitored only at the level of the protein product a special difficulty arises. For, even by detecting the appearance of a new protein soon after its synthesis, we cannot be sure that the genetic mechanism which controls its production does not operate well in advance of its appearance, even at a different stage of the cell cycle. This ambiguity has allowed biologists to adopt rather definite opinions on the question before us, often based on slender evidence either way.

With these cautionary remarks in mind, let us try to answer the question about alterations in gene expression with or without DNA

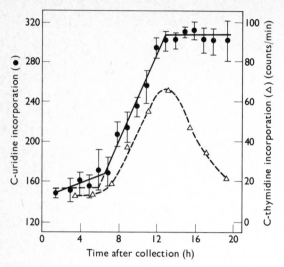

Fig. 2.13   DNA and RNA synthesis during the cell cycle. Rate of synthesis of RNA through the cell cycle of HeLa S_3 cells are shown by solid line and circles from pulses (10–30 min) of labelled uridine. The vertical bars are 95 per cent confidence limits. The dashed line and triangles are the rate of DNA synthesis from pulses of labelled thymidine. (After Pfeiffer S. E. and Tolmach, L. J. 1968. *J. Cell. Physiol.*, **71**, 77–93. Wistar Press.)

synthesis and/or cell division. As far as the bacteria are concerned, there can be little doubt about the answer.

Certainty in this area stems from the brilliant analyses of lactose fermentation in the bacterium *Escherichia coli* by Jacob and Monod and their colleagues. Their conclusions in regard to operator and regulator genes are more appropriate to question 4 in this chapter, and will be discussed there, but we should take note here of the important phenomenon upon which most of their researches depended. This is termed enzyme induction. In 1953 Monod and Cohen-Bazire published their observations that the enzyme β-galactosidase was not synthesized if *E. coli* were grown on media which did not require the utilization of this enzyme. However, within minutes of exposure to inducer (which was either the metabolite lactose or compounds of very similar structure) the enzyme was synthesized in quantity. An equivalent result was obtained with the enzyme penicillinase (Citri and Pollock, 1966). Early theories that induction was engineered by specific derepression of the gene coding the enzyme have now been largely confirmed, with the additional sophistication that an operator gene is involved in the derepression process. We can conclude that, in bacteria, inducible enzymes, such as penicillinase and galactosidase, can be induced very rapidly and apparently independently of DNA synthesis. (The level of inducibility actually doubles after DNA replication.)

There is also growing evidence that the levels of most bacterial enzymes are subject to regulation (Donachie and Masters, 1969) but the mechanisms involved remain obscure. However, it is certain that bacteria do alter their genetic expression without resort to DNA synthesis or cell division, and at least some of these alterations involve activation or repression of the genes.

In considering eukaryotic organisms, the first admission must be that clear answers to our question are hard to find. Far and away the most studied organisms in these terms are the yeasts, where our knowledge of gene expression and the cell cycle is now considerable (Mitchison, 1971, and Halvorson, Carter and Tauro, 1971). Here the differences in enzyme levels between the repressed and fully induced state of genes are much less dramatic than those found in bacteria. Often the basal level is high as compared with the few molecules of β-galactosidase to be found in an uninduced *E. coli* cell, and the levels after induction are often only some 25 times that of the uninduced state, which compares with the 1000 fold or higher increases of enzymes in bacteria following induction. Another feature of enzyme levels in yeasts is that the cell cycle imposes periods of restriction on the synthesis of some enzymes. In answer to the main question, alterations in enzyme levels occur in yeast cells without DNA synthesis, but whether these changes depend on actual gene repression or activation is not clear.

Turning now to other eukaryotic cells, the picture becomes more clouded. It is apparent that bacterial and yeast cells offer advantages for study which are lacking in the more complex plants and animals and it is not possible to review all the work which has a bearing on this subject, but three systems will be selected which seem to have provided the clearest answers.

Larvae of the two winged flies, *Diptera*, possess extremely large chromosomes in the cells of the midgut epithelium, the salivary glands, the malpighian tubule, the foot pad initials and the ovarian nurse tissue. These giant chromosomes are polytene, that is each chromosome has undergone repeated replication without separation of the strand copies. In addition, homologous chromosomes are intimately associated in an exaggerated form of somatic pairing. Up to several thousand chromatids thus lie side by side to give a structure of greatly increased breadth. Since the many chromatid strands in each polytene chromosome represent homologous structures, it can be assumed that the genes along their length will be in register, and when these chromosomes were first discovered, cytologists did notice that each had a curious and characteristic banded appearance. Was it possible that these bands were, in fact, the genes? Recent research suggests that bands may indeed include the functional units, and that the bands and interbands represent different states of chromatin organization (Fig. 2.14).

Unlike normal chromosomes which are only evident during nuclear division, giant polytene chromosomes are visible throughout the life of

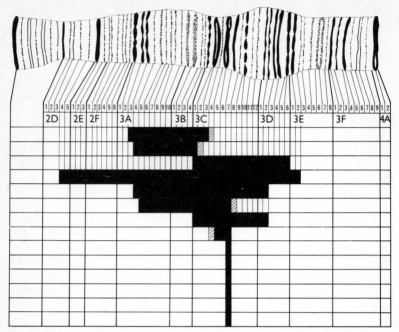

Fig. 2.14 A genetic map of the X chromosome of *Drosophila melanogaster*, matched to its banded structure. The black segments indicate bands removed in known deletions. Correlation of absent bands with altered phenotype is made possible by overlapping deletions. (After Swanson, G. P., Merz, T. and Young, W. 1967. *Cytogenetics*. Prentice Hall Inc.)

the cell. Is there then the hope of visualizing active genes on these chromosomes? The bands of polytene chromosomes can be shown to exist in two states—dense and chromatic, or swollen and diffuse. These swollen structures have been termed puffs (Fig. 2.15). Could puffs represent active genes? If salivary gland tissue of *Drosophila* is exposed to tritiated uridine, and an autoradiograph is made of the squashed cells, concentrations of label are found over the puffs, indicating that they are sites of active RNA synthesis. Presumably the puffs are regions in which the thousand or so strands of DNA making up the polytene chromosomes are opened out to render transcription of the local gene possible. Puffs, then, are genes visualized in action (Berendes, 1971). Actually, in some cases, more than one chromosome band appears to be involved in a puffed region, but good correlation exists between a specific puffed locus in the fly *Chironomus pallidivittatus*, and the appearance of a particular polypeptide in the salivary gland. Both are absent in the related species *C. tentans* (Grossbach, 1968).

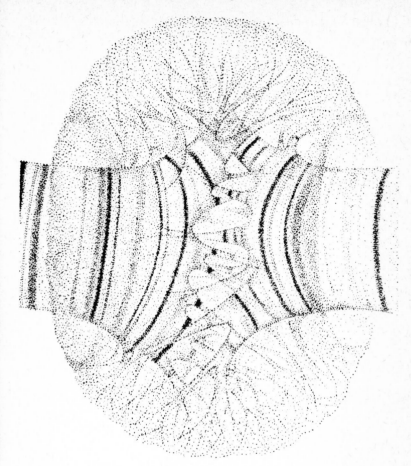

Fig. 2.15 Structure of a large puff on a giant polytene chromosome. (After Bullough, W. S., 1967. *The Evolution of Differentiation.* Academic Press Inc.)

Since puffs are active genes, then the polytene chromosome will surely provide an answer to our question about changes in gene expression without DNA synthesis. Do polytene chromosomes show alterations of puffing pattern without the intervention of DNA synthesis? The answer is that they do. As is discussed more fully in Chapter 6 on 'The Role of Hormones', the insect steroid hormone ecdysone can be used in an experimental system to induce changes in the *Drosophila* tissues. One of the responses to ecdysone is an alteration in puffing pattern. Moreover, these changes in pattern vary for different tissues but follow the same sequence for different cells within the same tissue, which is just what we would expect from a visualization of a changing genetic expression.

We must not conclude from these observations that this property of changing gene expression without DNA synthesis is a feature of all cells; but it is safe to assert that it is a feature of some. Dipteran polytene chromosomes offer many advantages in terms of size, persistence through the cell cycle, and altered expression under hormonal influence, and so it is not surprising that in no other cellular system is the picture as clear.

The mouse mammary gland is a particularly useful tissue for the study of gene expression since the synthesis of the milk protein casein can be assayed in cultures of mouse mammary gland, and the role of the critical hormones investigated. With this approach Turkington (1971) has shown that casein synthesis in mammary tissue is strongly influenced by three hormones, insulin, prolactin and hydrocortisone, and that when these hormones are provided, explants of mouse mammary gland undergo differentiation *in vitro*. Moreover, the explanted cells become organized into orderly nests of cells surrounding alveoli in a manner rather similar to that occurring in the normal tissue *in vivo*. Following these morphological changes, the synthesis of the milk protein casein and the enzyme lactose synthetase, both of which are unique to the mammary tissue, can be detected. With this model of differentiation in the test tube, Turkington has been able to prevent DNA synthesis and cell division and examine the influence of such inhibition on the differentiation process and synthesis of the milk proteins.

In this system, the differentiation of the milk producing epithelial cells occurs under the influence of insulin and hydrocortisone, and the subsequent production of casein is dependent on the presence of prolactin. If prolactin is withheld, very little or no casein is synthesized. If either DNA synthesis or further cell division is inhibited then casein synthesis fails even in the presence of optimal levels of prolactin. As shown in Table 2.2, the synthesis of casein can be prevented in these cells by suppressing DNA synthesis with lithium ions or hydroxyurea, while cell division can be stopped by the use of the mitotic inhibitor colchicine. Of course, we are here confronted with an experiment which

Table 2.2 Effect of lithium ion upon hormone-dependent synthesis of casein in isolated mouse mammary gland (Turkington, 1968)

System	Casein, cpm/mg tissue
Initial period	271
IFP	1081
IFP + Li$^+$, 20 mM	262
IF	263
IF + Li$^+$, 20 mM	278

Rates of casein synthesis were determined by exposing explants incubated in the indicated systems to ^{32}P during 32–36 h of incubation. I, insulin; F, hydrocortisone; P, prolactin.

monitors altered gene expression at the level of the protein product, which, as I have commented previously, is not ideal. Turkington has gone some way towards anticipating this objection by showing, in nucleic acid hybridization studies, that novel species of RNA are normally present in the casein synthesizing cell, and that these RNA species are undetectable in the tissues prior to this differentiation.

Although mammary gland activity is perhaps the best example, there are many other cases where DNA synthesis and/or cell division is needed for altered gene expression. For example, a system which closely parallels that of the mammary gland is the mammalian erythropoietic tissue which is responsible for the synthesis of the haem and globin, the structural constituents of haemoglobin. As with mammary gland, a hormone, appropriately termed erythropoietin, has a powerful regulatory role in the differentiation of this tissue. Although the details of the effect of erythropoietin are possibly less clear than in the case of prolactin, there seems no doubt that the stimulation of both haem synthesis and haemoglobin synthesis in bone marrow cell cultures is a property of the hormone. Erythropoietin also stimulates the production and differentiation of the erythroid cells. If DNA-synthesis is inhibited during the erythropoietin treatment, the hormone fails to elicit its response, again suggesting that the altered genetic expression required is precluded by either lack of DNA synthesis or prevention of division. The best evidence supporting this statement comes from work on the mouse (Cole and Paul, 1966); other workers have, however, recently shown that dependence on DNA synthesis does not apply to the effects of the hormone in mouse foetal erythroid cells (Marks and Rifkind, 1972).

Another situation in which DNA synthesis appears to be a prerequisite for altered gene activity is one of the few clear cases of enzyme induction in eukaryotes. From the work of Knutsen (1965) on the induction of the enzyme nitrite reductase by nitrite in the fresh water alga *Chlorella*, it is clear that induction is dependent on DNA synthesis, and the appearance of the enzyme is sensitive to both chloramphenicol and actinomycin D.

The last example to be discussed in detail provides strong evidence for a change in gene expression without DNA synthesis. Few, if any, cells reach a state of such complete genetic shutdown as the hen erythrocyte, and its normal level of RNA synthesis is low (Madgwick, Maclean and Baynes, 1972). No genes besides those for ribosomal and transfer RNA and some globin messenger RNA are likely to be transcribed in this cell. Using a technique pioneered by Okada, Harris and Watkins (1965) made artificial fusions between cells of different tissues and species, by exposing the cells to Sendai virus which had previously been inactivated by exposure to ultraviolet light. Various types of multinucleate fusion products were obtained; those products having nuclei of different kinds were termed heterokaryons.

Perhaps the most significant heterokaryons made by Harris were those which combined the nuclei of hen erythrocytes with nuclei from rabbit macrophages, mouse fibroblasts or HeLa cells (Harris, 1970). Two hen specific proteins were synthesized by such heterokaryons; the one was a surface antigen, the other a soluble enzyme, inosinic acid pyrophosphorylase, responsible for catalysing the condensation of hypoxanthine with ribosyl phosphate prior to its incorporation into nucleic acid. In this latter case a strain of mouse cells lacking the enzyme was used (Harris and Cook, 1969). Neither of these proteins is thus normally synthesized by the non-avian cell and their appearance in the heterokaryon is of considerable interest. Although neither of these proteins is made until the hen erythrocyte nucleus has developed a nucleolus, their synthesis appears to be independent of DNA synthesis. Harris and Watkins assumed that the chicken erythrocyte nucleus made no RNA. This is contrary to my own experience of chicken red cells

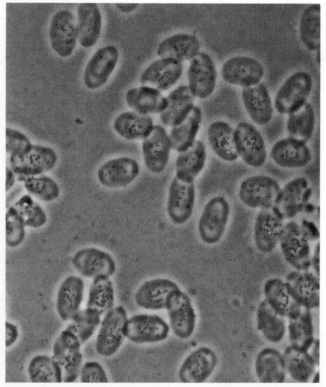

Fig. 2.16   Autoradiograph, showing grains over the nuclei of chicken erythro-cytes, indicative of RNA synthesis in these cells. The erythrocytes were washed and cultured *in vitro* for 5 hours in a medium containing tritiated uridine.

(Madgwick, Maclean and Baynes, 1972; see Fig. 2.16). While this reduces the weight of some of their evidence for altered gene expression in the fused cell their experiments give strong grounds for supposing that an alteration in genetic commitment and RNA synthesis may occur while a nucleus is still in the G_1 phase of the cell cycle.

How can these various threads be drawn together to provide an answer to the question? In some cells, bacteria and yeasts and insect salivary gland, there is no doubt that the pattern of genes being expressed can alter without recourse to DNA synthesis. This also appears to be the case for certain other cells, such as the chicken erythrocyte, although the evidence here is drawn from a highly artificial situation. Finally, in many tissues, including mammary gland and erythropoietic tissue, suppression of DNA replication or even a block to mitosis will prevent a hormonally induced change in the genes transcribed or the proteins made.

It has always been tempting to think that changes in the pattern of transcription might require a round of DNA replication, since at that time all genes are exposed to the DNA synthetic enzymes. What better time to adjust the mechanism of partial genetic shutdown? But it is not so simple. Perhaps, as Gurdon and Woodland (1968) have suggested, major changes in expression do occur only at that time, and mammary gland differentiation may be an example of this, but it is certain that the generalization is so open to qualification that it is scarcely worth making. Many cells *do* alter their gene expression without either DNA synthesis or mitosis.

2.4 If changes in gene expression occur between divisions, how are they controlled? Such changes are normally viewed at the level of the protein product. Are these changes transcriptional or post-transcriptional?

So far in this chapter we have talked as if transcription of a gene into RNA, and translation of that RNA message into protein, were so closely related in time that they could usefully be treated as one event. This is very far from being the case, and we must now address ourselves to the problem of their precise relationship.

The reason why gene expression is so often discussed at the level of the protein product is one of pure expediency. There is often no alternative. Proteins betray their presence by their enzymic action, their colour (in the case of haemoglobin), or by their specific antigenicity. So it is possible to discern the pattern of proteins in a cell and detect changes in that pattern during differentiation. But for every protein made, there is at least one RNA intermediate molecule, the so-called messenger RNA. Obviously gene expression should be monitored at the level of the messenger RNA molecule and changes in expression detected as changes in the spectrum of these molecules in the cell. Unfortunately, techniques for isolating and identifying these molecules are still in their infancy and only in such naturally fortuitous cases as the polytene chromosome or repetitious genes can gene action be assayed at that level.

How serious is the fault of measuring gene activity and modulation purely at the protein level? At least in bacteria little is lost. To judge from the speed of appearance of the new protein in cases of enzyme induction, only a few minutes elapse between activation of the gene and detection of that event by an assay for the enzyme. The example of 'zygotic induction' is even clearer. Here a new gene is introduced by bacterial conjugation into a cellular environment in which it can be expressed. Zygotic induction is normally used to describe lysis of a zygote bacterium following mating between a male Hfr bacterium possessing an integrated prophage in its genome, and a female F − recipient bacterium which lacks the lysogenic characteristic (lysogenic bacteria are not lysed by the phage). Within minutes of the integrated phage DNA entering the sensitive bacterial cell, synthesis of phage protein can be detected. Shortly afterwards the sensitive cells are lysed and the newly formed phage released.

Moreover, most bacterial protein synthesis is highly sensitive to the suppression of RNA synthesis by Actinomycin D, and the now classic experiments of Gros *et al.* (1961) and Brenner, Jacob and Meselson (1961) depended for their successful diagnosis of the role of messenger RNA, on the short time lag between RNA synthesis on the gene and protein synthesis on that RNA.

Whenever attention is shifted from prokaryotic to eukaryotic cells, the relationship between gene transcription and translation becomes more complex.

To give a correct impression of the temporal relationship between eukaryotic transcription and translation in a few words is impossible, but there are two techniques which have contributed more than any other to resolving this problem, and we will briefly indicate the main conclusion which derive from these methods. The techniques are (a) the use of Actinomycin D, which inhibits gene-determined RNA synthesis, and (b) enucleation, which effectively removes all nuclear genes and therefore the principal source of cellular RNA. Both techniques have some drawbacks. Actinomycin D affects the synthesis of some types of RNA (particularly ribosomal) much more than others, and it may have side effects on the permeability of the cell membrane. Removal of the cell nucleus is compatible with survival only in a few cell types which may be unusual or atypical in some sense. We must therefore exercise caution in the interpretation of data from each type of experiment.

In yeast cells the experiments of Mitchison and Creanor (1969) reveal that, for the enzymes sucrase and alkaline phosphatase, even the protein-synthesis inhibitor cycloheximide does not accomplish an immediate reaction. Another system which reveals the complex nature of the connection between gene transcription and translation in eukaryotes is that of enzyme synthesis in rat hepatoma cells, utilized by Peterkofsky and Tomkins (1967). Their work demonstrates that, when treated with Actinomycin D, the induction of tyrosine aminotransferase was affected, but the persistent synthesis of other enzymes was not. These observations hold good for most eukaryotic cells, where the synthesis of different proteins is affected in a highly variable fashion by Actinomycin D.

In view of its alleged lack of RNA synthesis, the chicken erythrocyte has often been quoted as an example of a cell in which synthesis of the protein haemoglobin was dependent on long lived messenger RNA. Actually this cell does synthesize some RNA, and when Actinomycin D is included with a culture of these cells in a test tube, it has a substantial effect on the level of protein synthesis (Maclean and Madgwick, 1973). Even in this cell, then, translation is far from being independent of transcription.

If this observation is compared with the finding that the mammalian reticulocyte, devoid of its nucleus, will continue to synthesize haemoglobin for some days, it will give a fair impression of the variable nature of the dependence of protein synthesis on gene activity even for one type of protein in different organisms.

Cell enucleation experiments are certainly responsible for impressing on biologists the frequent independence of translation and transcription in eukaryotes. The large unicellular marine alga, *Acetabularia*, has proved particularly suitable for such experiments. Scattered information from some other enucleated cells is also available. The most crucial evidence emerging from studies on *Acetabularia* is that *many weeks* after removal of the cell nucleus, differentiation of the characteristic cap of the alga can be accomplished (Hammerling, 1963; see also p. 103). Moreover, there is no doubt that cap formation requires the synthesis of new and specific enzymes and a fairly major change in the synthetic commitment of the cell. Since *Acetabularia* is a plant cell with both chloroplasts and mitochondria, considerable amounts of DNA already exist in the cytoplasm and perhaps this or other cytoplasmic DNA fractions is responsible for encoding the information for the cap protein. Such an argument is met by the finding that the morphology of the cap is determined by the nucleus and *not* by the cytoplasm. This discovery was made by exchanging nuclei between species of *Acetabularia* in which the form of the cap was conspicuously different. The form of the cap is dependent on the species origin of the nucleus, not that of the cytoplasm. We can conclude from the *Acetabularia* experiments that, at least in some eukaryotic cells, gene transcription and RNA translation *may* be separated by a long period of time.

No other enucleated cells continue to survive for anything like as long as *Acetabularia*. Work on *Amoeba proteus* by Ord (1968) shows that many amoebae survive for over five days after removal of the nucleus by microsurgery, while enucleate *Stentor* cells can survive for about 4 days and enucleate fragments of mammalian tissue cells for about the same time. The nearest approach to *Acetabularia* seems to be made by the freshwater alga *Spirogyra*, in which cells without nuclei have survived for about 2 months in the laboratory of Dr. J. Hammerling (1959).

In the face of the long survival of cells lacking nuclear DNA and the resistance of the synthesis of many proteins to Actinomycin D, we are forced to admit that measuring gene activity at the level of the protein product is unsatisfactory. Moreover, the problem does not end there, for the now substantial data in favour of a post-transcriptional selection of species of RNA for translation tells us that it is not only a time lapse which separates the two levels. Although the primary control of gene expression is therefore transcriptional, there may be many other factors determining the final expression in terms of an assay of completed protein or a complete differentiated cell. The varieties of control can conveniently be considered at three levels:

(a) Transcription,
(b) Post-transcriptional (up to, and including translation), and
(c) Post-translational.

A consideration of these three levels of cellular control will help to answer the final question about the mechanism of genetic expression, and this forms the basis of the next chapter.

3

Control mechanisms in gene expression

3.1 Transcriptional control and alterations in gene frequency

In the first section of this chapter, we will discuss not only the regulation of the process by which RNA is produced, but any control at the level of the DNA. This will include mechanisms which increase or reduce the amount of the DNA and those by which the DNA available for transcription is regulated and limited: I have discussed the topic of the control of gene expression at some length in another book (Maclean, 1976).

(a) Alteration in chromosome number

The crudest level of genetic control is the elimination from a cell, or a line of cells, of whole segments of chromosomal DNA. Indeed, the most surprising conclusion to emerge from the early studies on differentiation was that the genetic control of differentiation did not, in general, operate in this way. The identity of DNA composition and chromosome complement for differing tissues in the same organism must surely have been initially unexpected. But perhaps we have been so schooled into an acceptance of this dogma that we ignore the many exceptions to it. These we will now briefly review.

In many groups of simple plants, for example the mosses, most of the life cycle is haploid and the diploid stage of the organism, produced by the fusion of the gametes, is soon followed by a meiotic division and a return to the haploid state. Bacteria are also haploid for much of their cell cycle. In most animals, on the other hand, the adult is diploid and only the germ cells are haploid. A particularly interesting exploitation of haploid/diploid variation occurs in many Hymenopteron insects, including the honeybee *Apis*. In the honeybee, not only are the sperm haploid, but all males develop from unfertilized haploid eggs. Similar systems occur in some other Hymenoptera. In these cases, male haploidy results from parthenogenetic development of unfertilized eggs, while females develop from fertilized eggs. Parallel cases of parthenogenesis are to be found amongst rotifiers, mites and many insects.

Many organisms possess cells with an altered ploidy. Mammals have polyploid cells in liver tissue, while in invertebrates many somatic cells are regularly polyploid. The same is true of plants (see Fig. 3.1). For example, plants of the family Leguminosae possess tetraploid cells in the tissues of their root nodules.

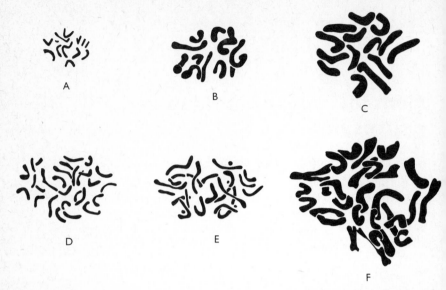

Fig. 3.1 Squash preparations of chromosomes from leguminous plants showing the increase of chromosome number in the cells of the root nodules. (A) red clover, ordinary root tip cell; $2n = 14$. (D) nodule tissue of red clover with 28 chromosomes. (B) common vetch, ordinary root tip cell; $2n = 12$. (E) nodule of common vetch with 24 chromosomes. (C) garden pea, ordinary root tip cell; $2n = 14$. (F) nodule of garden pea with 28 chromosomes. (After Srb, A. M. and Owen, R. D., 1955. *General Genetics*. H. Freeman & Co.)

Cells or organisms deficient or duplicated for a particular chromosome are termed aneuploid, and sex determination in many organisms involves aneuploidy. Thus in many insects the female possesses two X chromosomes but the male has only one, and there is no Y chromosome (Lewis and John, 1968).

This is an interesting contrast to the mode of sex determination in most mammals where the female diploid cells possess two X chromosomes and the male an X and a Y. Since the monosomic state of a single X chromosome being present without a Y leads to the development of a clinical female phenotype (Turner's syndrome), we can conclude that the presence or absence of the Y chromosome acts as a sex determining device.

We have now catalogued a few examples in which variation in the overall chromosome complement, or of any one chromosome, is associated with the differentiation between sexes or castes of the same species or between different tissues in the same organism. We can now turn to examples where actual chromosome elimination from the cell is associated with tissue differentiation. It is perhaps remarkable that this has not proved a more frequent method of regulating genetic expression but, in fact, there are not many examples. Two are particularly well known. One is to be found in *Parascaris equorum*, a parasitic nematode worm. This creature has only two very large 'compound chromosomes' in the zygote, and these break up in certain of the cleavage mitoses to yield numerous small chromosomes. At the same time there is a loss of the terminal portions of the compound elements. The pattern of diminution and loss is such that only the germ line retains the two 'compound chromosomes' intact (Fig. 3.2). So here then, is an example of differentiation involving genetic reduction in some tissues as compared to others.

The other rather well documented case of chromosome elimination is found in some insects, especially a genus of fungus gnats (*Sciara*) and the family of the gall midges (*Cecidomyiidae*), (Fig. 3.3). In both these cases,

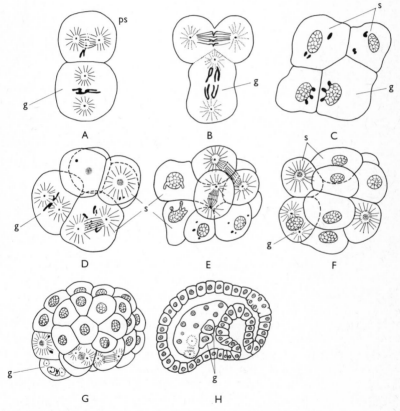

Fig. 3.2 Cells of developing embryo of *Ascaris megalocephala*, showing chromosome diminution in the somatic cells. The two large chromosomes are retained in the germ cells (marked g), while dividing somatic cells (s) reject some of the small chromosome pieces at mitosis. (After Wilson, E.B. 1925. *The Cell in Development and Heredity*, Macmillan, New York.)

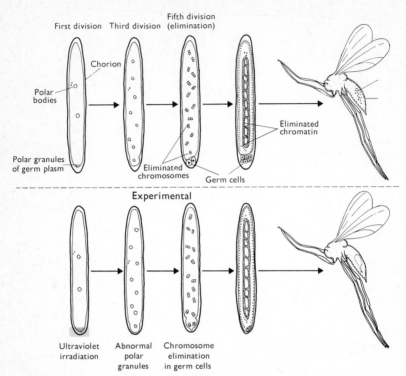

Fig. 3.3   The process of germ cell formation in the gall midge *Myetiola* in normal and experimental conditions. After ultra-violet radiation the germ cell nuclei are not protected from chromosome elimination and the result is a sterile fly. (After Bullough, W.S. *The Evolution of Differentiation.* 1967. Academic Press.)

as in *Parascaris*, some chromosomes are eliminated from the somatic cell lines while being retained in the germ line. In both these insect groups the situation is complex and further information can be found in *Cytogenetics* by Swanson, Metz and Young (1967).

All these examples of increasing or reducing the total gene pool of a line of cells are presumably devices aimed at altering a tissue by having some genes silenced by their absence or other genes emphasized by their multiple presence. But there are other ways of achieving these goals and, indeed, differentiation by increasing or decreasing the amount of genetic material is the exception, not the rule. Much more frequently genes are silenced by suppression or emphasized by an increase in the number of copies of the gene actually being transcribed. So we can study transcriptional control in its more conventional sense in the ways by which cells permit some genes to be transcribed while, at the same time, retaining those that are not.

(b) Selective transcription of available genes

Assuming that, with the exceptions discussed above, different tissues in the same organism have an identical genome, the transcription of these genes could be affected at two levels. In the one, whole blocks of genes or indeed entire chromosomes could be suppressed while others remained active. At the other level suppression would operate selectively on single genes or small groups of genes. Both levels of transcriptional control occur in nature and we will begin by looking at the first, the temporary or permanent suppression of whole blocks of linked genes. A fuller account of chromatin activation and repression will be found in a recent review (Maclean and Hilder, 1976).

When chromosomes are studied in stained preparation certain of them, or in some cases parts of them, are regularly found to stain more densely than others and also to stain earlier in the division cycle. These regions are said to be heteropycnotic (different in density), and are relatively constant in their behaviour. For this reason they have been described by the term *heterochromatin*; that chromatin which does not show this intense and early staining being distinguished as *euchromatin*. After the first genetic maps had been constructed, it became clear that these regions of heterochromatin were either genetically inert or inactive—that is few or no genes on the map appeared to be within the areas of heterochromatin. So heterochromatin gained the reputation of representing that chromatin which was inactive within the cell. Unfortunately, the term heterochromatin is now often used more widely for locally inactive chromatin, whether it can be recognized as heteropycnotic in cytological preparations or not (Frenster, 1969). This is, of course, somewhat confusing and we will reserve the use of the word heterochromatin in this book for the narrower original application.

A most interesting example of the heterochromatinization of a chromosome occurs in most somatic cells of the human female and affects one of the two X chromosomes. Somatic interphase nuclei from a normal human female contain a dark Barr body on one side of the nucleus. This is a heterochromatic X chromosome. In females with an abnormal karyotype consisting of three or one X chromosomes, the number of Barr bodies is proportionately adjusted and is always one less than the total X chromosome complement (Fig. 3.4). According to the now widely accepted hypothesis of Dr. Mary Lyon, this represents a mechanism of 'dosage compensation', whereby in female somatic cells, as in male, only one X chromosome is functional.

The two X-chromosomes of a human female are of distinct parental origin. It is not difficult to demonstrate that heterochromatinization is random with respect to the two X-chromosomes since X-linked loci will, when heterozygous, lead to a mosaic state, some cells expressing the normal allele and others the mutant. Thus the gene responsible for the formation of the enzyme glucose-6-phosphate dehydrogenase is X-linked. Beutler *et al.* (1962) were able to show that red blood cells from

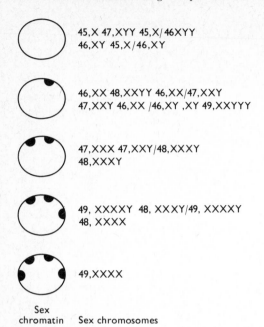

45,X 47,XYY 45,X/46XYY
46,XY 45,X/46,XY

46,XX 48,XXYY 46,XX/47,XXY
47,XXY 46,XX /46,XY ,XY 49,XXYYY

47,XXX 47,XXY/48,XXXY
48,XXXY

49, XXXXY 48, XXXY/49, XXXXY
48, XXXX

49,XXXX

Sex
chromatin   Sex chromosomes

Fig. 3.4 Schematic diagram to show the correlation between the number of X chromatin masses (Barr bodies) and number of X chromosomes. (After Hamerton, J. L., 1971. *Human Cytogenetics*. New York, Academic Press.)

women heterozygous for this gene were of two types, some carrying normal enzyme levels and some totally deficient for the enzyme. We can also surmise, then, that apparently normal women who, as carriers of haemophilia, are heterozygous for that gene, are in fact mosaics for the expression of that locus, but the produce of the cell population expressing the wild type allele is sufficient to ensure entirely efficient clotting of the blood cells.

It is well to note in passing that heterochromatic chromosomes are not always entirely genetically inactive. The human sex chromosomes illustrate this well, since XO individuals are abnormal females (Turner's syndrome).

In addition to these cases where entire chromosomes are heterochromatic, others exist where only particular areas show heteropycnosis. The most obvious and important is the centromeric area of many chromosomes. This we will discuss further in the section on gene reiteration and amplification.

Turning to the question of how heterochromatin is inactivated, we do well to ask what is known of the chemical and structural differences between heterochromatin and euchromatin. The short answer is that, with the exception of centromeric heterochromatin already mentioned, we know very little. What we can enumerate are the observed characteristics of heterochromatin. They are the more dense staining already alluded to, a later onset and a faster speed of DNA replication than euchromatin, and an absence of RNA synthesis. Under some circumstances areas of chromatin normally classed as euchromatin display some or all of these characteristics. It is not surprising, therefore, to find that the two states are, in many cases, reversible or at least that one set of chromosomes may be euchromatic in one cell and heterochromatic in another—for example, the inactivated X chromosome which forms the Barr body. The paternally derived set of chromosomes of the mealy bugs (coccids) become heterochromatic in the male but remain euchromatic in the female (Brown and Nur, 1964). Such cases are examples of facultative heterochromatin as opposed to those where a given chromosome or chromosome region is heteropycnotic at all times and in all tissues (constitutive heterochromatin, see p. 84). The most likely explanation for the heteropycnosis of heterochromatin is that the chromatin fibres are more tightly packed together than they are in euchromatin. Assuming that the differences between the two types of chromatin are essentially explained by condensation and packing, then it is relevant to look at other sources of condensed and dispersed chromatin. Frenster (1969) has succeeded in isolating fractions of calf thymus chromatin which conform to these categories, but again, has so far been unable to detect real biochemical differences in structure.

The only conclusion possible is that heterochromatin is differently organized and more highly condensed than euchromatin, but the factors which determine these differences are not at present understood. It is also highly probable that the large areas of heterochromatin which we can so easily see in Barr bodies and B-chromosomes are structurally similar to much more confined areas of heterochromatin not normally detectable by staining techniques, but distributed throughout the length of many chromosomes (Maclean and Hilder, 1976). We should also not ignore the possibility that the observed condensation of heterochromatin may often be secondary to its inactivity. Cells showing no inactivated X chromosome as a Barr body are often found, nevertheless, to have a genetically silent X chromosome and there is some evidence for a general disparity of base composition between the DNA of euchromatin and heterochromatin (Comings, 1972). In general, while we can ask penetrating questions about the regulation of gene expression at particular gene loci, we remain largely ignorant about the broadest and most obvious kind of transcriptional control at the level of whole chromosomes or regions of chromosomes.

The cell system which provides the clearest picture of transcriptional control at the level of single genes is undoubtedly that of prokaryotes. Of course they offer many advantages for the investigator, such as their easy culture in immense numbers, the phenomenon of enzyme induction, and the lack of basic protein bound to their DNA. Since the publication of Jacob and Monod's classic paper in 1961 the concept of the operon has dominated all discussions of transcriptional control (Fig. 3.5). In essence, the mechanism of the operon is that *cistrons* (i.e. genes coding for single polypeptides) for enzymes involved in a common pathway are situated together on the polynucleotide strand and are controlled as a unit. Control of their transcription is the responsibility of the operator gene, which is also located near to the group of cistrons. The operator gene plus the relevant cistrons is called the *operon* in Jacob and Monod's terminology. By a comparative study of the behaviour of a series of bacterial mutants, these scientists discovered that the operator gene was sensitive to the product of another gene, which they termed the regulator gene (I) situated in a neighbouring part of the bacterial DNA. Their explanation of induced enzyme synthesis is that, in the uninduced state, the regulator gene produces a repressor molecule specific for the operator gene. When the repressor molecule is situated on the operator gene the whole operon is turned off and none of the cistrons are transcribed. Inducer molecules, they suggest, combine with the product of the regulator gene and, perhaps by an allosteric effect (Fig. 3.6), prevent its effective repression of the operator gene. As a result of this they turn on the operon and allow transcription of the appropriate cistrons.

Present understanding of the galactosidase operon in *E. coli* has been complicated by the discovery of a gene located beside the operator gene (o). This gene, of about one hundred nucleotides in length, has the role of initiating transcription by the polymerase enzyme into messenger RNA. Such a role was originally allocated to the o gene itself. The new gene, now recognized as controlling the start of transcription, has been termed the promoter gene (p). At first, prior to 1968, the promoter gene was believed to lie between the o and Z loci, but there is now little doubt that p precedes o in the operon and thus the order of the lac operon is I–p–o–Z–Y–A, with a space between the I and p genes. Here:

I = regulator gene   Z = β-galactosidase enzyme gene
p = promoter gene   Y = β-galactosidase permease gene
o = operator gene   A = β-galactosidase transacetylase gene

By careful recovery of inducer molecules from within recently induced bacteria we might expect to gain a clue about the identity of the repressor molecule, the product of the regulator gene (I). Such experiments confirm that the repressor molecule is a protein. The repressor of the *lac* operon in *E. coli* has been isolated and is a tetrameric protein (Gilbert and Muller-Hill, 1966). It is also interesting to note here that a protein has been isolated which binds cyclic AMP and, when thus conjugated, increases the efficiency of the promoter gene although binding to a separate site in the promoter region to that occupied initially by the RNA polymerase molecule.

In addition to such systems of induced synthesis, dependent on de-repression of the appropriate genes by the inducer, bacteria possess

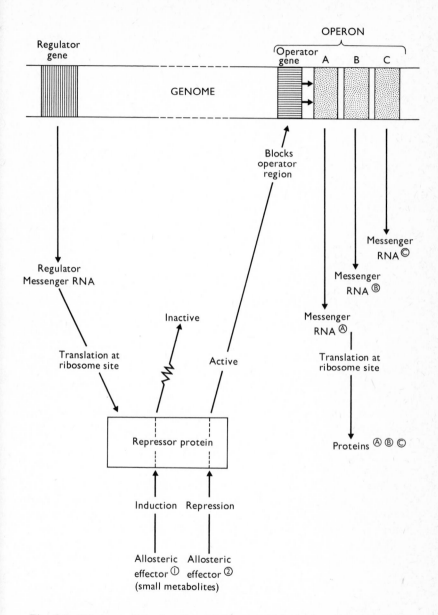

Fig. 3.5 Diagram of the Jacob and Monod model of gene regulation in the lactose operon. The promoter region is not shown. (A, B and C represent three structural genes.) (After Whittaker, J.R. 1968. *Cellular Differentiation*. Dickenson Publishing Co., Inc., California.)

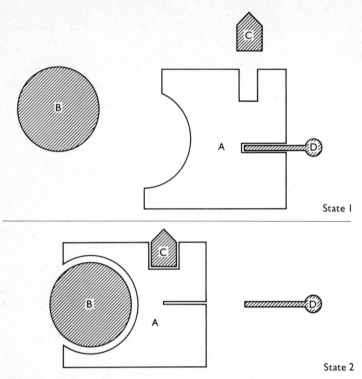

Fig. 3.6 *Diagram to illustrate allostery.* In state (1), the molecule A is not coupled with molecule B at the appropriate attachment site, thus permitting attachment of molecule D but excluding C. However, in state (2), on attachment of molecule B, the shape of A changes, thus excluding D but permitting attachment of C.

other transcriptional control systems operating positively rather than negatively. In these, the operon will only function when switched on by an inducer which is itself the product of the regulator gene. Here control is exerted in an opposite way to the 'lac' operon, since the operon is repressed by the binding of an 'external' repressor molecule to the inducer molecule thus rendering it unfit to induce the inactive operon. L-arabinose metabolism in *E. coli* almost certainly provides an example of this control mechanism and once more the active product of the regulator gene is a protein.

Given that bacterial transcription involves such positive and negative regulation, we can proceed to enquire what proportion of their genome is governed in this way. The answer is that, although some systems—for example the enzymes involved in arginine synthesis—have characteristics which do not fit easily into any operon-like category, most bacterial cistrons are controlled by induction and repression involving both controlling genes and regulatory molecules, some of the latter being essentially intracellular, others being 'effector' molecules which enter the cell from the external medium.

When we transfer our attention to studies on eukaryotic transcriptional control, we are immediately faced with a much more confusing and poorly understood area. Various workers, inspired by the beautiful experiments and clear conclusions associated with the bacterial work, have made brave attempts to apply this information and understanding to eukaryotic cells. But it is not easy. Let us begin by reminding ourselves of some of the general characteristics of eukaryotes and their genetic material: (a) the principal genetic material is partitioned from the cytoplasm by nuclear membranes (b) the genetic material is invariably strongly associated with histones and other proteins, (c) the sheer amount of the DNA is much greater than in bacteria, and often apparently in vast excess of any genetic requirements, (d) differentiation built on multicellularity means that many or most cells carry genes unlikely to be expressed at any time in the life of that cell.

Even brief consideration of these generalized characteristics must assure us that the simplistic genetic control of bacteria will not do for higher cells. Of course, even on evolutionary grounds, we might expect some operon-like systems to have survived in eukaryotic cells, and this does appear to be the case. Gene clusters for certain pathways have been identified in some fungi. The galactose system in yeast displays some characteristics of an operon, but some other groups of enzymes, originally cited as evidence for operon control, are now thought to be examples of the translation of an intact polycistronic messenger, and no relevant operator loci have been detected.

How, then, is genetic transcription controlled in eukaryotic cells? The most likely mechanism is surely within the chromatin itself. Since most of the non-DNA material present in chromatin is the basic protein histone, this seemed at first the most likely substance responsible for gene regulation (Fig. 3.7).

However, despite years of wishful thinking and wishful experimentation, the fact is that histones only partly fill the early expectations of this role. The first disappointment to these early ideas was the gradual realization that not only was the overall histone content of heterochromatin and euchromatin essentially the same, but that even the different species of histone were present in the two types of chromatin in much the same ratios. And what obtains for hetero- and eu-chromatin also applies to the chromatin extracted from different tissues within the same organism. If histones really do fulfil the key role of selective gene repression in the eukaryotic cell, the least that we can expect is that different differentiated tissues from the same organism will have different amounts of histone or different types of histone on their chromatin. A large body of evidence now points the other way. Histone is quantitatively and qualitatively similar in widely different tissues within the

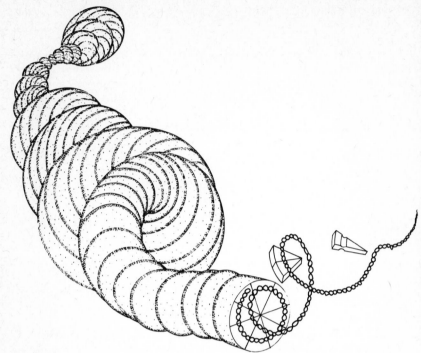

Fig. 3.7 Hypothetical structure of a chromosome fiber. Inside each chromosomal fiber, a single DNA double helix is tightly packed, first by supercoiling to form a type A fiber and then by supercoiling again to form a type B fiber. The DNA molecule is held in its super-coiled configuration by histones and other DNA linked proteins, which are shown hypothetically as wedge-shaped molecules with two binding sites each. (After DuPraw, E. J. *DNA and Chromosomes*, 1970. Holt, Rinehart and Winston, Inc., New York.) This model of DNA/histone association is now rendered unlikely by the evidence that DNA itself is looped around beads of aggregated histone (Baldwin *et al.*, 1975).

same organism (see Fig. 3.8). It is also rather similar even when extracted from unrelated organisms. One fraction, known as histone IV, is almost identical whether derived from calf thymus or pea plants. This may also be interpreted as implying that all of the sequence is important in function and not just the DNA binding site (Comings, 1972).

This is not to say that histone is not *involved* in gene regulation. What it does imply is that histone is not itself the specific gene-recognizing molecule. It would indeed be surprising if histone, a chief constituent of chromatin, had no role to play in genetic regulation. Why else would it be there? Admittedly other functions of stabilizing the DNA and protecting it from undue degradation or alteration have been ascribed to histone, but they are scarcely adequate in themselves. Moreover,

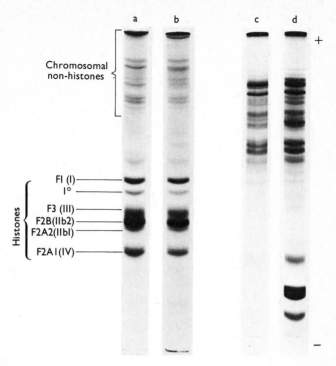

Fig. 3.8 Electrophoresis of chromatin proteins. Proteins extracted with urea NaCl from (a) cerebellum and (b) pituitary. Note the identity of the histone bands. (c) and (d) are gel separations of cerebellar chromatin protein, non-histones prepared by SDS extraction in (c) and total chromatin protein after prolonged urea NaCl extraction in (d). (From Shaw, L. M. J. and Huang, R. C. C. 1970. *Biochemistry*, **9**, 23, 4530–42, Fig. 5. The American Chemical Society.)

histone does regulate gene expression in a rather general way. It also seems possible that cyclic AMP may be involved in the phosphorylation of histones (Langan, 1968). If pure DNA and histone-complexed DNA are compared for RNA synthetic ability in the test tube, there is no doubt that the histone-complexed DNA shows a severely reduced availability for transcription into RNA. Some early and clever experiments of James Bonner, described in his book *The Molecular Biology of Development* (1965), makes this point clear. Bonner extracted DNA and chromatin from peas and pea embryos and studied their ability to support RNA and protein synthesis. In particular, he also added histone to the pure DNA and found that gene transcription could be greatly reduced in this way. We would now conclude, however, that his experiments involving the synthesis of specific proteins require a more complicated explanation than the one which he originally inferred. Actually DNA continues to act as a good template even when complexed 1:1 with basic proteins and it may be that histones act in a rather general way by inhibiting the extensive progress of polymerase enzymes (Koslov and Georgiev, 1970).

Chromatin contains molecules besides DNA and histone; in particular small amounts of RNA and acid protein are present. Both of these last named molecules are possible candidates for the role of regulators of specific gene function, and, to date, both molecules have been backed for this role by different workers in this field. The most plausible model is that involving the acid protein, and Dr. John Paul and his collaborators in Glasgow have accumulated strong evidence in favour of what they term 'organ specific gene masking' by acid proteins (Paul, 1971) although they warn us that, even in their system, the preparations of acid protein are contaminated by 1 or 2% RNA. It is interesting to recall here that the specific repressor molecules which have been isolated from bacteria have proved to be acidic proteins. The model proposed by Paul's group is very similar to that offered by Frenster in 1965. As illustrated in Fig. 3.9, it supposes that histone acts as a general transcription repressor, and that the genes which are active in any particular tissue have an altered relationship with the histone. This altered relationship, Paul suggests, is the result of interference by gene specific acid protein, which can displace the histone by offering to it alternative anionic binding sites.

Let us be well aware of the uncertainties in Paul's model. Firstly, the active specific unmasking agent may not be the acid protein, but rather the few per cent of RNA contaminating the acid protein preparations. Huang and Huang (1969) have certainly taken this view. Alternatively it might be a ribonucleoprotein. Secondly, some weaknesses can be found in Paul's detection system. The procedure adopted was to make RNA on templates of DNA or chromatin, using bacterial RNA polymerase enzyme, and then to hybridize these RNA fractions in competitive conditions against DNA from the same organism. Such an approach might be expected to indicate the similarity or divergence between different RNA samples by determining whether they competed for the same DNA sites or hybridized satisfactorily with alternative sites. Paul has himself listed a number of weaknesses inherent in this experimental approach, such as the likelihood of relatively non-specific hybridization in conditions of saturation, the possible infidelity of reconstruction of chromatin from DNA histone and acid protein, and the somewhat unnatural transcriptional conditions introduced by the use of exogenous polymerase. The greatest restriction placed on any general application of these studies to eukaryotic transcriptional control stemmed from the fact that the hybridization condition detected only highly repetitive DNA sequences, and so what was being measured was the tissue specific differences in availability of these sequences. We will discuss such DNA sequences and their significance later in this chapter.

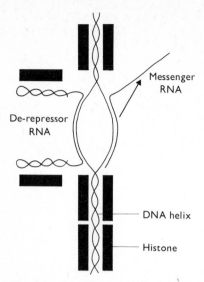

Fig. 3.9 Frenster's model for selective gene transcription. Polycationic histone repressors interact electrostatically with the negative phosphate groups on the exterior of the DNA helix, stabilizing the helix against strand separation and consequent transcription. Nuclear polyanions form complexes with such histones nonselectively, partially displacing these histones from the underlying DNA helix, permitting accelerated acetylation of the N-terminal groups of the displaced histones, and allowing separation of the strands of the DNA helix in a random fashion. De-repressor RNA species, specific for particular gene loci, hybridize with the anti-coding strand of DNA within such loci, and stabilize the strand separation loop in the open position. The remaining (coding) DNA strand within the stabilized separation loop is then free to serve as a template for gene-specific and strand-specific asymmetric transcription. (After Frenster, J. H. 1965. *Nature Land,* **206**, No. 4990, 1269.)

Actually this objection has been largely met by the experiments of Barrett *et al.* (1974), which point strongly towards acidic proteins acting as specific gene regulators in eukaryotic chromatin. I have discussed the probable role of non histone protein in eukaryotic transcriptions more fully elsewhere (Maclean and Hilder, 1976 and Maclean, 1976) and for the moment we must be content to conclude that there is some evidence that either acidic proteins or RNA may play a key function of regulating specific gene transcription in eukaryotic cells. It should also be made clear that much of the acid protein extracted from chromatin may be structural or enzymatic rather than regulatory. Pea bud and calf thymus have broad similarities when compared at the level of non-histone-histone proteins (Elgin and Bonner, 1970).

More speculative models of eukaryotic transcriptional control have been advanced by Georgiev (1969) and Britten and Davidson (1969) both stimulated by the apparently excessive amounts of DNA in the genome of most eukaryotes. Both of these models suppose that numerous genes in the genome are purely regulatory in function and are probably never translated into protein. Such genes, termed sensor, integrative, receptor and producer genes in the terminology of Britten and Davidson, would help to account for the abundance of DNA in higher cells. These models also provide a role for the very high molecular weight RNA to be found in the nucleus and the observed repetitious sequences on the DNA. The model of Maclean and Hilder (1976) is a third alternative which stresses the possible role of DNA and protein in the controlled condensation of the chromatin. It is a sharp reminder of our ignorance of eukaryotic gene regulation that we can neither prove nor disprove the general application of any of these theories, and that they owe more to our understanding of prokaryotic rather than eukaryotic transcriptional control.

We have talked so far as if genes were distributed entirely at random within the genome and that, bacterial operons apart, regulation of gene expression requires recognition of the gene not by its site but by its base sequence. In fact, there are at least some exceptions to this assumption and it may be that the assumption is largely untrue. The genes coding for protein in the tails of T (even) phages are in two clusters: since at least 21 such genes have been identified, this system also suggests that many genes related to the organization of a structure may be regulative in function (Levine, 1969). In bacteria, for example, there are many well authenticated examples of genes coding for enzymes in the same pathway being clustered close together on the genome. Not all of these clusters are functional operons, and gene clustering is not confined to bacteria; it has been reported in the fungus *Aspergillus nidulans* (Cove, Pateman and Rever, 1964) in yeast, (Douglas and Hawthorne, 1964) and in *Drosophila*, (Lewis, 1964). To what extent these are all special cases we do not known, but we should not exclude the possibility that some clustering of functionally related genes persists in eukaryotes. An observation not unrelated to gene clustering is that of 'position effect', a phenomenon by which the position of a gene on the genome affects its expression. One of the best documented cases of position effect stems from experiments on eye colour in *Drosophila* (Becker, 1966 and see Fig. 3.10). A mutation called white (w), which is a recessive allele of the locus responsible for red eyes (w^+) is normally located on a euchromatic portion of the X chromosome. If in w^+/w heterozygotes the w^+ locus is moved by structural chromosome rearrangement to the proximity of heterochromatin, then flies with mottled or variegated eyes are produced. This results from the inactivation of w^+ and shows that the position of a gene in the genome can affect expression. Cases are also known where the gene most affected by the 'position effect' is not the one closest to the heterochromatin (Lewis, 1950).

Whether gene clustering and position effects are highly unusual in

Through appropriate crosses, flies with various chromosome constitutions can be produced. For example :

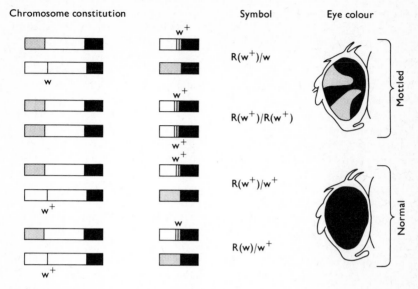

Open bars: euchromatin of chromosome I
Black bars: heterochromatin
Grey bars: euchromatin of chromosome IV

Fig. 3.10 Position effect variegation at the white locus in *Drosophila*. (From Markert, C. L. and Ursprung, H., 1971. *Developmental Genetics*. Prentice Hall Inc.)

eukaryote genetics, or whether they betray distinctive and fundamental aspects of the genetic organization of higher cells is not known. But we would be well advised to remember that gene position can affect its expression and this suggests that position may affect the *timing* of gene transcription.

Having suggested that gene position might affect the time of gene transcription, we can approach the matter from the other end and ask whether there is any evidence that gene read-out is influenced by gene order. Since the genetic maps of most eukaryotes remain sketchy, this question is not easy to answer. However, one type of cell in which the time of appearance of enzyme during the cell cycle can be monitored, is yeast. Two conclusions emerge from studies on enzyme synthesis in growing yeast cells. The first is that particular enzymes appear at particular times during the cell cycle, and the second is that the temporal sequence of synthesis of different enzymes is related to their position along a chromosome. There is somewhat less evidence for the second conclusion than the first, but it is such an interesting idea that any evidence in its favour deserves close scrutiny. This work on yeast cells and the conclusions emerging from it have been well reviewed by Halvorson, Carter and Tauro (1971). If some form of temporal order on the chromosome is a device used to regulate the time of gene transcription, then the gradual differentiation and maturation of, say, a red blood cell, could be partly engineered by a sequential reading along its DNA, genes coding for early enzymes being read early and the globin genes read late. Also as pointed out by Ashworth (1973, p. 21), there is probably close coupling between the cell cycle regulation and cell differentiation in eukaryotic cells.

Often in scientific research, there are situations of special significance, or difficulty, which pose challenges to the inventors of general theories. There is one such biological situation for theorists on eukaryotic transcriptional control. It is the genetic regulation of antibody synthesis. Antibodies are, of course, fairly complicated molecules with an important immunological role in vertebrate animals. Each antibody consists of two small (light) polypeptide chains and two larger (heavy) chains, ultimately held together by disulphide bonds and operating as an integrated unit (Fig. 3.11). The peculiar problem of the genetic control of antibody structure is that both light and heavy chains are made up of two distinct regions. Each chain has a so-called C (constant) region, which varies only a little and is obviously, for many different antibodies, the product of the same gene, and a V (variable) region, which varies greatly and is equally obviously, for many different antibodies, the product of different genes.

How can we explain the common contribution of one cistron to the products of many other cistrons in the sequential synthesis of polypeptide chains? The evidence that the C regions of many varied antibodies in one organism are the products of one gene behaving in a normal Mendelian fashion is absolutely clear, and so too is the evidence that multiple genes are involved in encoding information for the multiple forms of the V region; yet deletions are known which involve adjacent stretches of V and C regions. Assuming that any one lymphoid cell makes only one type of antibody, and that the variation of antibody structure is genetically determined (see Burnet, *The Clonal Selection Theory of Acquired Immunity*, 1959) it seems inescapable that the two genes responsible for the V and C region of any one antibody chain within a cell must operate as a unit despite their original separate locations within the genome. Moreover, there are grounds for thinking that the actual number of cistrons for the V region is substantially less than the number of variable types of V chain, and that some of the V region variability is the result of somatic translocation within the V region cistrons. Before we leave the subject of antibody gene control, we should mention another phenomenon which has been revealed by studies in this area. This is the process of 'allelic exclusion', by which only one of two allelic forms of an immunoglobin chain is expressed in any one cell. No precise parallel for this observation is known outside immunoglobulin

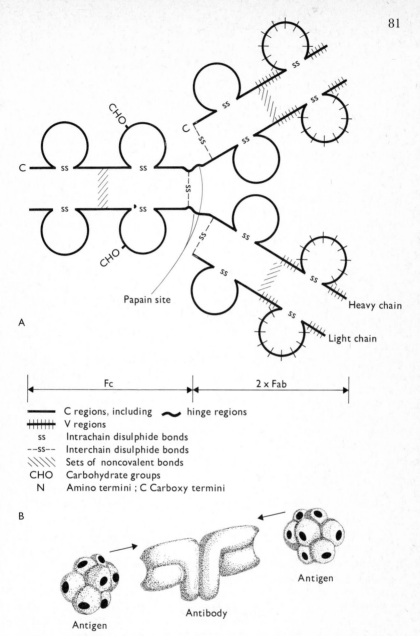

A

| Fc | 2 × Fab |

——— C regions, including 〜 hinge regions
┼┼┼┼┼ V regions
ss Intrachain disulphide bonds
--ss-- Interchain disulphide bonds
\\\\\ Sets of noncovalent bonds
CHO Carbohydrate groups
N Amino termini ; C Carboxy termini

B

Antigen

Antibody

Antigen

Fig. 3.11 (A) Diagrammatic representation of an antibody molecule. (B) Model to show the binding of antibody to two antigens. Existence of two identical antigen-combining sites on each 7S antibody molecule. Both light and heavy chain atoms are used to form the combining sites. (A. Diagram kindly provided by Prof. G. Stevenson. B. After Watson, J. D., 1970. *Molecular Biology of the Gene*. W. A. Benjamin Inc.)

synthesis though it is certainly reminiscent of the X-inactivation found in female mammals (see p. 67). The sickle cell allele, for example, does not show allelic exclusion in the heterozygous condition, since both normal and sickle cell beta globin chains occur in the same cells.

(c) Repeated and amplified genes

Our discussion of transcriptional regulation has so far assumed that each gene is present within the haploid genome as only one copy, and indeed until recent years this was held to be generally the case. We now know that there are many exceptions to this generalization and these we must now discuss.

One technique is almost solely responsible for revealing the existence of multiple gene sequences and providing information about their significance. This technique is known as nucleic acid hybridization, and by its use complementary strands of nucleic acid, whether DNA to DNA or DNA to RNA, can be matched up. An important contribution has also been made by the use of density gradient centrifugation to separate satellite bands of nucleic acids. Salts such as caesium chloride, when in solution, have densities which cover the range of densities of DNA of differing base composition (1.6–1.8 g cm^{-3}) and centrifugation to equilibrium in these solutions yields good separation of differing DNA fractions. Nucleic acid hybridization has been applied both with isolated preparations of nucleic acid, and *in situ* on the chromatin of chromosomes within cells. Detailed information about the technique can be found in the reviews of Walker (1971) and Flamm (1972).

It is perhaps useful to outline briefly the basis and *modus operandi* of nucleic acid hybridization. In such experiments, a sample of DNA is mechanically sheared so that the molecules are approximately the same size, and then the sample is 'melted' or partly denatured by heating under carefully controlled conditions. The 'melted' nucleic acid is now single rather than double stranded—that is the hydrogen bonds linking the double helix have broken and each polynucleotide chain is now a separate molecule. Since both viscosity and ultraviolet absorption at 260 nm increase when DNA is melted, these parameters can be used to monitor the procedure. If the melted sample is now incubated at a temperature below the melting temperature, provided that such factors as molecular weight of the DNA and ionic strength of the media are optimal, annealing or renaturation of the DNA will occur, complementary base pairs will once more bond and the double helix will zip-up. The actual rate of this annealing process can be used as a measure of the base sequence complementarity of the DNA sample, and single DNA strands present as multiple copies will obviously anneal more quickly, since perhaps millions of correct complementary strands will be available.

The usual way of expressing the annealing process is to plot the percentage of annealing against the C_0t, where C_0 is the original DNA concentration and t the annealing time (in seconds). Figure 3.12 shows a typical $C_o t$ curve for mouse DNA and the analysis of frequency of sequence types which can be decuced from it.

Problems still exist with the technique, and particularly with repeated sequences, some 'fold back' hybridization may occur even at very low $C_o t$ values by the folding over and annealing of part of a strand with its own repeated sequences elsewhere in the strand. The discovery that many sequences exist in a palindromic arrangement (reading the same in two directions from an initiation point) may permit fold back even with

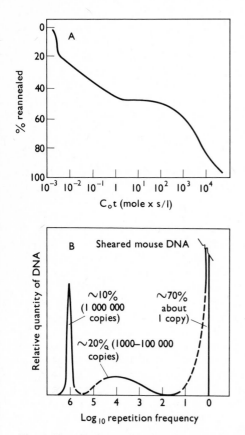

Fig. 3.12 (A) Curve showing the rate of reannealing of mouse DNA. Such a $C_o t$ curve yields information about the composition of the DNA sample. (After Ashworth, J. M., 1973. *Cell Differentiation.* (B) Graph representing the frequency of repetition of nucleotide sequences in mouse DNA (dashed segments represent regions of uncertainty). This graph is derived from such curves as are shown in A. (After Britten, R. J. and Kohne, D. E., *Science,* **161**, 529–540. Copyright 1968 by The American Association for the Advancement of Science.)

so called unique sequences. Palindromic sequences may also help to explain the observation of bidirectional replication of DNA polymerase enzymes (Callan, 1973).

Although initially most laboratories employing the technique of nucleic acid hybridization hoped to compare the genomes of both closely and distantly related species, a strange and unexpected fact gradually emerged. This was that most organisms possessed a fraction of DNA which hybridized much faster than any other and that early comparisons between DNA of different species had been mainly a comparison between these confined parts of the genome. It became clear that the greater the number of copies of a given nucleotide sequence within a single genome, the more rapidly would that fraction hybridize. It also became apparent that the DNA involved in this rapid hybridization consisted of many thousands of copies of closely similar and very short sequences. We cannot adequately discuss here either the complicated experiments which have utilized nucleic acid hybridization or the pros and cons of the possible conclusions which emerge from them. What we will attempt is a modern view of the eukaryotic genome in the light of all of these studies.

The genetic material of the eukaryotic cell must now be thought of as comprising rather distinct and separate regions and although the evidence is drawn from the DNA of only a few types of cells, there are no apparent reasons for doubting its general application.

(i) *Highly repetitive sequences*

This type of DNA is sometimes confusingly referred to as nuclear satellite DNA. The percentage of the genome occupied by this material varies between 1 and 60% and it consists of several million copies of short lengths of some ten to twenty nucleotides. Many of these short lengths are identical, but others show slight variation. The nuclear satellite DNA of the guinea pig consists of very many repeated copies of the simple sequence $\frac{CCCTAA}{GGGATT}$ (Southern, 1970). Of particular interest is the finding that, for even closely related species, the amount of this type of DNA is very variable, a fact which had led many people to surmise that it has a high rate of evolutionary turnover. By hybridizing very highly radioactive satellite DNA (the highly repetitive fraction) to chromosomes within fixed cells, Pardue and Gall (1970) and Jones (1970) were able to trace the chromosomal location of this material in the mouse. They have shown that most of it is in the centromeric area with some additional concentration in the nucleolus organizer. This suggests that, whatever its other functions, the highly repetitive DNA may be involved in the movements and attachment of chromosomes to one another and to the spindle. In the kangaroo rat (*Dipodomys ordii*), Bostock *et al.* (1972) find that 60% of the entire genome is DNA of this type, that it replicates exclusively in the second half of the S phase of the cell cycle, and that the short arms of many of the chromosomes in this species consist of this type of DNA.

(ii) *Ribosomal and Transfer RNA Genes*

Insects, amphibians, birds and mammals possess between 100 and 1000 copies of the genes for 28S and 18S ribosomal RNA clustered together. Mycoplasma possesses a single ribosomal cistron and bacteria between 5 and 10 copies. The actual repeated sequence is made up of one DNA segment for 28S, one for 18S, a spacer RNA portion which includes a 5.8S RNA sequence, and a 'non-transcribed' spacer region (Birnstiel, Chipchase and Speirs, 1970 and Speirs and Birnstiel, 1974). Some outstanding electron microscopy has provided visual evidence of this organization (see Fig. 3.13). Except in bacteria ribosomal 5S RNA is not coded within the 28S and 18S ribosomal gene cluster. But clusters of ribosomal 5S RNA are spread throughout the genome, often, apparently, concentrated at the ends of chromosomes. A forfuitous biological discovery helps us to be fairly sure that only one cluster of 28 and 18S ribosomal genes is usual in the haploid genomes of eukaryotes, namely the occurrence of a mutant form of *Xenopus laevis* with a chromosomal deletion of that area. Tadpoles with only one nucleolus per cell were discovered, and when the resulting adults were crossbred, a proportion of the larval offspring proved to possess no nucleoli and to be unable to synthesise 28S and 18S ribosomal RNA.

Transfer RNA sequences are also repeated, probably some thousands of times and, like the 5S genes, appear to occur as a series of separate clusters.

The phenomenon of gene amplification, which applies to the ribosomal cistrons, we will discuss separately later in this chapter (see p. 95).

(iii) *Other repeated sequences*

Aside from the highly repetitive DNA sequences associated with the centromere and the nucleolus, and the genes coding for ribosomal and transfer RNA, there is evidence for other sequences being present as multiple copies. From the data obtained with hybridization studies (Hennig and Walker, 1970) it appears that about 20–30% of a eukaryotic genome consists of repeated sequences. Much of this DNA occurs as families of short related sequences, too short to function as genes for polypeptide synthesis. Such sequences also seem to be distributed widely throughout the eukaryotic genome, and may serve as control elements (Firtel and Lodish, 1973, and Davidson *et al.*, 1973).

Environmentally induced genetic changes. For some years now it has been known that apparently permanent changes in certain varieties of plants, particularly flax, could be induced by environmental changes, such as growth in higher temperatures and increased fertilizer levels (Durrant, 1962). It now appears that these changes also involve increases in the

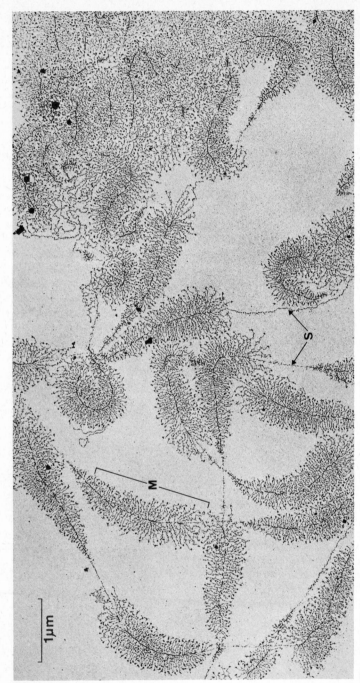

Fig. 3.13 Electron micrograph of amphibian nucleolar DNA showing details of the RNA and protein containing matrix. The feather-like profiles are stages in the synthesis of the ribosomal RNA, as in section M, while S represents a spacer region. (Courtesy of O. L. Miller and B. R. Beatty. 1969. *J. Cell. Physiol.*, **74**, No. 2. part III, 225–32.)

amounts of certain repeated sequence DNA within the genomes of the affected plants (Cullis, 1973).

(iv) *Unique sequence DNA*

Largely arising from the use of the hybridization technique, most of the cellular DNA has come to be referred to as unique sequence DNA, and is that part of the genome which contains most of the genes coding for proteins, and may be present in non-repeated or slightly repeated copies. At the present time the hybridization technique cannot normally be relied on to distinguish between one and perhaps twenty copies. Thus the term unique is somewhat inappropriate. This is particularly obvious when we discuss the possible existence of multiple copies of 'unique sequences'.

'Unique sequences' (multiple copies). From investigations so far undertaken there is little evidence for such multiple copies. The protein which seems the most likely candidate for dependence on multiple gene copies is globin, since the erythroid cells make this protein in very substantial amounts. However, although there is still some conflict of evidence on the matter, globin genes do not appear to be highly repeated, and there may be as few as one copy per globin species (Bishop, Pemberton and Baglioni, 1972). Histone is another class of proteins made by most cells in substantial quantities, and here the evidence is more favourable, since Kedes and Birnstiel (1971) report a high level of gene repetition and clustering.

There are two alternative hypotheses for interpreting the role of repeated copies of normal genes. One is to assume that they might carry out some function, regulatory perhaps, other than coding for cellular protein. Another is that such repeated sequences are accidental. Since DNA synthesis is now known to commence with a small section of RNA synthesis, (Sugino, Hirose and Okayaki, 1972), one wonders if a gene is ever replicated by mistake instead of being transcribed to yield messenger RNA molecule. If it were, the resulting gene might either be incorporated in tandem next to the original, or incorporated elsewhere in the manner of SV40 viral genome insertion, after its release from the chromosome (Maclean, 1973). Yet another alternative hypothesis is a variant of one which we have already discussed, that is the necessity for multiple copies of many cistrons. We must concede that there is little enough information to help us choose between these alternatives, but the concept of multiple gene copies deserves a little more discussion because of the important consequence it has for our model of the eukaryotic genome. A major objection to the existence of multiple copies of cistrons for cellular proteins is that any protein coded by numerous identical gene copies would be extremely resistant to mutational change. No such proteins are known. Thus, in the case of human globins, the cistrons for α-, β-, and γ-globin do mutate readily, as is evidenced by the many abnormal haemoglobins such as the sickle cell haemoglobin which carries an altered amino acid in the chain. Can we have it both ways? Attempts to construct a model of the eukaryotic genome which permits conventional mutation *and* multiple gene copies are not new, and most fall into a framework known as the *master-slave* hypothesis originally proposed by Professor H. G. Callan of St. Andrews, Scotland (Callan, 1967) to explain certain features of amphibian lampbrush chromosomes. A somewhat elaborated version of the original hypothesis has been suggested by Whitehouse (1967). Callan's hypothesis proposes that the multiple copies of any one gene consist of one 'master' copy and numerous 'slave' copies, and the 'slaves' are regularly matched against the 'master' sequence, presumably by repair enzymes, so that mutations in the 'slaves' would be automatically corrected but mutation of the 'master' would be passed on to all the 'slave' copies. Perhaps the greatest attraction of the 'master-slave' hypothesis is that it would explain some puzzling observations emerging from other studies. For example:

(a) Synthesis of RNA along the loops of amphibian lampbrush chromosomes occurs in a pattern which suggests that the DNA is being unwound at one end of the loop, spun in at the other and transcribed only when exposed in the loop (Fig. 3.14). Callan has interpreted this to mean that each loop consists of numerous slave copies of the same gene.

(b) The genomes of many or most eukaryotes are organized into chromomeres; these are especially evident in the giant polytene chromosomes of *Diptera*, where homologous chromeres aggregate as bands. Chromomeres are much too large to be single genes or even operons, and yet genetic markers in *Drosophila* are, to date, roughly equivalent to one per chromomere.

(c) Some Dipterans vary by having 2, 4 or 8 times as much DNA in some chromosome bands as is present in other closely related subspecies, and yet the chromosomes do not vary in polyteny between the two subspecies (Fig. 3.15).

(d) The amounts of DNA present in the diploid cells of amphibians and species of *Lilium* are extravagantly high (Fig. 3.16) and, in at least some of these—for example, the amphibian *Amphiuma*—a large percentage of the DNA hybridizes as if it were repetitious.

(e) Some molecular evidence can be drawn from the electron microscope studies of Thomas *et al.* (1970), in which DNA circles can be formed from a large proportion of eukaryotic DNA, suggesting tandemly repeated nucleotide sequences. Such sequences were not restricted to regions of specific DNA density suggesting a rather general distribution throughout the genome.

(f) The 'spacer' sequences of the ribosomal cistrons of *Xenopus laevis* and *Xenopus mulleri* have been compared. There are about 700 identical sequences in each of these ribosomal clusters, and so different are the sequences between these two species that no

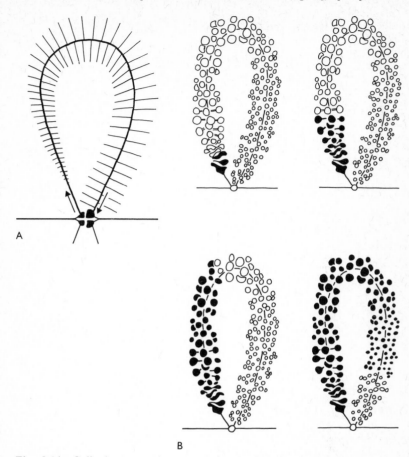

Fig. 3.14 Callan's proposed structure for the lampbrush chromosome. (A) Diagram showing the 4-part structure of a lampbrush chromosome and the postulated movement of the loop in a polarized direction. (B) Sequential labelling of the giant granular loop on chromosome XII after being exposed to labelled RNA precursors. (After Callan, H. G. 1963. *Int. Rev. Cytol.*, **15**, 1–34. Academic Press, New York.)

homology is detectable. Yet, within one of these species, the 700 sequences display complete homology (D. D. Brown quoted in Flamm, 1972). How else but by a master-slave hypothesis could the intra species homology in spacer sequences be explained, if evolution can change them so radically between species?

(g) The many copies of the ribosomal and transfer RNA copies within a given organism appear to be extremely similar (Birnstiel *et al.*, 1968).

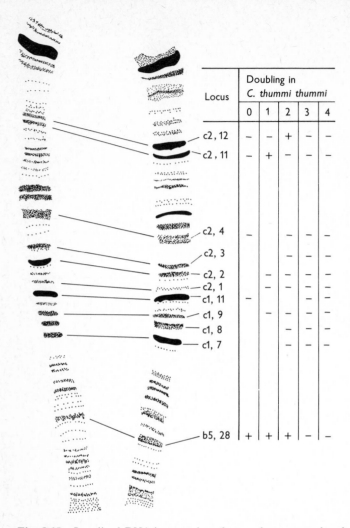

Fig. 3.15   Localized DNA increase in polytene chromosome bands. The mid-region of the right arm of chromosome 11 in the polytene nucleus of the salivary gland in the hybrid *C. thummi thummi* × *C. thummi piger*, showing the localized and geometric increase of DNA in certain bands. Right, *C. thummi thummi*; left, *C. thummi piger*. (After Keyl, H. G. *Chromosoma*, **17**, 139–80.)

(h) Numerous workers have reported the isolation of RNA of very high molecular weight from cell nuclei of many tissues and species. This RNA has often been termed heterogeneous or heterodisperse in that it does not sediment in a narrow band (Fig. 3.17). It is an attractive possibility that this RNA might be the

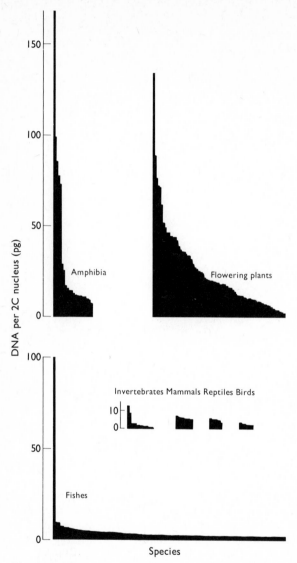

Fig. 3.16 The range of variation in nuclear DNA. Species in descending order of nuclear DNA amount. (After Rees, H. and Jones, R. N. 1972. *Int. Rev. Cytol.*, **32**, 53–92. Academic Press Inc.)

transcriptional product of the multiple gene complexes for many genes, and there is at least one bit of evidence supporting the idea that the globin mRNA sequence is included in this nuclear RNA fraction (Williamson, Drewienkiewicz and Paul, 1972).

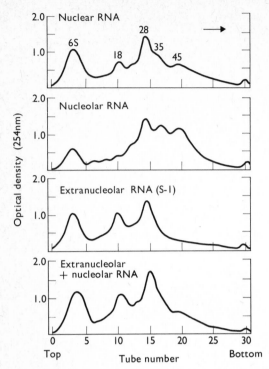

Fig. 3.17 Sucrose density gradient profiles of nuclear, nucleolar, and extra-nucleolar nuclear RNA of rat liver. RNA's prepared by an SDS-hot phenol extraction procedure are placed on a 10–40% sucrose gradient containing 0.1 M NaCl with 0.01 M Na acetate, pH 5.1, 1 mM EDTA, and centrifuged at 25,000 rpm for 17 hours. The bottom graph is the composite profile made by the sum of the nucleolar and extranucleolar nuclear RNA. (After Maramatsu, M. 1966. *Biochim. Biophys. Acta.*, **123**, 119, Fig. 1.)

(i) In studies on DNA replication in Amphibia, Callan (1972), fol-lowing on from experiments by Huberman and Riggs, has shown rather neatly that amphibians with widely differing amounts of DNA replicate their DNA in unit lengths in rough proportion to the overall DNA content (Fig. 3.18). Two ideas follow from this. One is that DNA synthesis, like RNA synthesis, occurs in distinct units on the genome, and these units may be the same for both kinds of synthesis: the other is that, where two related organisms have widely differing total DNA values, this may be largely a function of the multiplicity of gene copies.

(j) Eukaryotes have a lower deleterious mutation rate than do pro-karyotes. This would follow if many correctable slave genes occurred between master genes. Callan's ideas on master-slave models have been reviewed at length by Thomas (1970) and by Ris and Kubai (1970) and widely discussed by Comings (1972).

These observations which lend weight to a master slave model should not blind us to some strong evidence which argues against it. For example:

(a) Studies involving nucleic acid hybridization suggest that much of the mammalian genome exists as unique sequences, or at least sequences with very low multiplicity (Gall *et al.*, 1971).

(b) Hybridization of DNA from a single band of a *Drosophila* chromosome suggests unique sequence DNA, not multiple copies of one gene. (Ris and Kubai, 1970).

(c) There is evidence that the giant nuclear heterogeneous RNA is a precursor of a smaller, probably messenger RNA, and that much of the original molecule is not translated.

(d) Two mutants of the human globin genes, haemoglobin Constant Spring and haemoglobin Tak, involve addition of 31 and 10 amino acid residues onto the ends of the α and β chains respectively. In neither case do these additional sequences resemble the beginning or end sequences of the globins concerned.

(e) Recent evidence from a number of laboratories (Davidson *et al.*, 1973) suggests that unique sequences and repetitive sequences roughly alternate along the length of the DNA. But since the most pressing evidence for this particular picture comes from the slime mould *Dictyostelium* (Firtel and Lodish, 1973) which has a very low genome size, perhaps this may have only limited application to the normal eukaryotic situation.

Complementation groups. We must now introduce a recent term, the genetic complementation group, which is that length of DNA which includes all the regulatory genes and multiple slave-master cistrons for one protein product. It is likely, judging by the molecular weight of nuclear RNA and the DNA replication units observed in eukaryotic genetic organization that such complementation groups are a major feature of eukaryotic genetic organization. Whether or not the eukaryotic genome consists substantially of multiple gene copies we cannot, as yet, be certain, and it is arguable whether the model solves more problems than it poses. It would however provide a neat method of rate control at the transcriptional level and so provide a useful genetic device in cell differentiation.

For the moment it seems foolish to elaborate on this intriguing phenomenon until we learn more about it.

'Unique sequences' (single copies). Until the facts about multiple gene sequences are clearer, it is not possible to be certain about the proportion of DNA which is made up of single gene copies. But the mapping of the phage and bacterial genomes and the regulatory genes within them

94

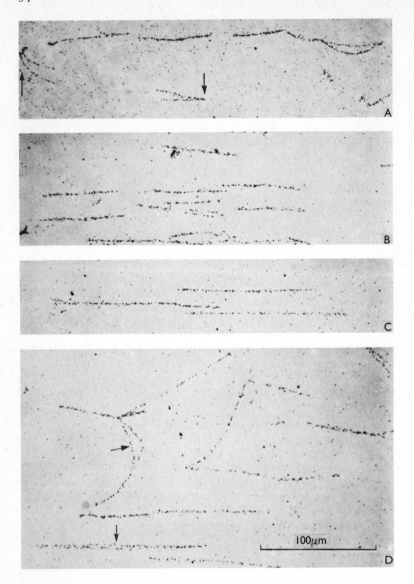

Fig. 3.18 Photographs of a DNA fiber autoradiograph originating from tissue culture cells grown in the presence of ^{3}H Thymidine. (A) is from a *Xenopus laevis* culture labelled for 4 h and exposed for 28 weeks. Arrows point to the divergent replicating forks of a segment which started replication before (^{3}H) thymidine was provided, hence the unlabelled gap, and in which sister double helices had evidently separated from one another when trapped on the underlying filter. (B) is from a *Triturus cristatus carnifex* culture labelled for 2 h. This auto-radiograph was exposed for 23 weeks. Notice how the labelled segments are at

certainly permit us to say that many genes in prokaryotes must be present as unique copies. Even the modest DNA content of most mammals is probably 1000 times greater than is needed to code for all cellular proteins if all sequences were unique: however, if a fair proportion can be allotted to repeated sequences and regulatory functions, perhaps we can assume that the residue is made up of unique gene copies.

It must now be clear that a really central problem to differentiation and indeed to the whole of biology, is how the expression of these unique sequences is controlled in eukaryotic cells. At present little is known. If these sequences do frequently alternate with groups of short repeated sequences (Davidson *et al.*, 1973), then perhaps these repeated regions act as promoter regions and govern the transcription of the cistron next to them. It would also be reasonable to speculate that 'tissue master' genes may exist, exerting control over many promotor regions and therefore over many different cistrons (Maclean and Hilder, 1976). Perhaps such 'tissue master' genes represent the expression of differentiation at the tissue level, so that there is one control gene for erythrocyte differentiation, one for neuron differentiation and so on, each of these central control genes therefore determining the activity of the battery of genes characteristic for a tissue or differentiated cell line. Such ideas have little or no evidence to support them and must await the results of future experiments for their vindication or discard.

(v) *Gene amplification*

In a few organisms early development makes exceptional demands on the supply of ribosomes, and a special regulatory device has been developed to meet such demands. This device involves the manufacture of very large numbers of copies of the 28S and 18S ribosomal RNA genes within the oocyte. The 5S cistrons are not amplified and nor are the genes coding for ribosomal proteins. The term *gene amplification* has been given to this phenomenon, and it should be retained for such use and not confused with the terms gene reiteration or repetition, which refer to repeated sequences normally present in the genome.

Of course these ribosomal genes are already present as multiple copies in the genome but amplification raises the number of copies of the ribosomal RNA transcription unit from about 1500 in the normal somatic cell of *Xenopus laevis* to about 2 million in the oocyte. This

---

least as long as those of *Xenopus laevis* when labelled for 4 h. (C and D) are from a *Triturus cristatus carnifex* culture labelled for 4 h. These autoradiographs were both exposed for 23 weeks. Long labelled segments, most of which do not show clear evidence of tandem arrangement, are characteristic of this *Triturus* material, indicating that initiation points for DNA replication are much further apart than those of *Xenopus*. In figure D arrows point to segments where sister double helices are separate. (From Callan, H. G. 1972. *Proc. R. Soc. B.*, **181**, 19–41. Figs. 16–19.)

phenomenon is obvious even in the light microscope since very large numbers of nucleoli, each containing at least one complete ribosomal gene cluster, develop within each oocyte nucleus. Gene amplification is an essentially transient process and has so far been observed in a limited number of eukaryotic species, such as *Xenopus laevis*, the house cricket *Acheta*, and the water beetle *Dytiscus*. Actually in both the insects mentioned, there is some suggestion that DNA other than that coding for the ribosomal cistrons, is also amplified, but most of it is heterochromatic and is not transcribed.

It has recently been suggested that gene amplification may be achieved via an RNA intermediate, utilizing the enzyme reverse transcriptase (Crippa and Tocchini–Valentini, 1971). At the moment of writing, however, evidence for the existence of this enzyme outside RNA virus replication processes is weak, and we must conclude that it is a possible but unsubstantiated mechanism.

The term gene amplification could be stretched to include the production of multiple copies of the mitochondrial and chloroplast DNA (see Chapter 4). However, the term is not normally used in this way. It may be that, in certain situations, genes other than the ribosomal genes are amplified, and some evidence for this has come from study of a mouse myeloma tumour (Krueger and McCarthy, 1970).

3.2 Post transcriptional control

The importance of post transcriptional control mechanisms is particularly difficult to discuss without having any clear ideas about the permanence or impermanence of the differentiated state. We cannot with certainty say that any differentiated state is permanent or even irreversible, and yet there seems small point in discussing changes in enzyme levels which are so impermanent as to be unworthy of the term differentiation. This difficulty is especially apparent when thinking about bacteria. In these cells there are mechanisms of negative feedback or end product inhibition, by which the end product of a synthetic pathway, say an amino acid, will effectively block the continued synthesis of that amino acid by acting directly on the enzymes operative in the early steps in the pathway. If a culture of bacteria is divided into two batches, and excess of one amino acid added to one batch, then the bacteria become metabolically different through the operation of such a post transcriptional control mechanism, though these differences scarcely qualify for the title of differentiated states. However, it is at least conceivable that such metabolic changes, occurring as they do in eukaryotic as well as prokaryotic cells, may in certain situations operate as differentiating mechanisms. At a crucial stage in development, access to exogenous amino acid, and consequent shut down of its endogenous synthetic pathway, might lead to profound and essentially permanent differences between tissues or organisms.

However, we will now discuss the more obviously differentiating mechanisms operating post transcriptionally: these are all processes which can alter or modify the transcribed message of the genes at some stage prior to or during its translation into protein. We have already taken note earlier in this book of the comparative inadequacy of monitoring altered gene expression at the level of the protein product, and an appreciation of post transcriptional control should deepen our awareness of this difficulty. The term translation control will be avoided, firstly because there is no evidence for control being exerted during the process of translation, and secondly because the term would be too narrow and might exclude control occurring after transcription but prior to actual translation.

(a) Nuclear processing of the products of transcription

Stemming from the initial observations of Professor Henry Harris in 1959, many laboratories have now established that there is a turnover of RNA within the nucleus of the eukaryotic cell. Not all the products of transcription leave the nucleus. While arguments about levels of nuclear protein synthesis persist, it seems fairly certain that much of the nuclear RNA which is turning over is never translated. It is not unreasonable to argue, then, that there is a screening mechanism which allows only those RNA molecules to leave the nucleus which are wanted for the translation process of that cell at that time or at some predictable future time. Studies on the RNA isolated from nuclei have now revealed that, aside from the transfer and ribosomal RNAs, and their precursors, nuclear RNA consists of extremely heterogeneous classes of molecules, many of which are apparently heavier, and perhaps much heavier, than the ribosomal RNA molecules. Much of this RNA never leaves the nucleus, or at least, it cannot be isolated from the cytoplasm in that state.

Figure 3.17 illustrates fairly typical sucrose gradient profiles of nuclear RNA. It is immediately obvious that this heterogenous (or heterodisperse) nuclear RNA must include messenger RNA. Sequences of SV40 virus RNA have been identified within the high molecular weight nuclear RNA of infected cells (Lindberg and Darnell, 1970), and reports from the laboratory of John Paul and elsewhere suggest that globin messenger sequences are to be found in the heavy nuclear RNA (Williamson, Drewienkiewicz and Paul, 1972). For the most part we can assume that, as with the ribosomal precursor RNA, the messenger molecules which are released into the cytoplasm form only a part of these larger molecules. The residue of the nuclear RNA is presumably destroyed as redundant. Of course there is absolutely no direct evidence for *selective* breakdown being used as a screening device, but we may take it that the mechanism involved in the processing of the ribosomal RNA precursor is a pointer in that direction. By this mechanism the immediate transcription product of the 28 and 18S ribosomal genes is 'cut down' to size by enzymic cleavage and reduction to yield single 28S and 18S molecules. It is probable, then that the first step in post-transcriptional control is selection of RNA molecules for release to the cytoplasmic translational machinery. This offers potentially powerful modification of the immediate expression of the nuclear genes.

Following its initial transcription, but prior to transport to the cytoplasm, a proportion of nuclear RNA is adenylated, that is, a 'tail' of some 200 adenine residues is attached to its 3′ end. Not all nuclear RNA receives such poly-A sequences, and there is some evidence to suggest that nuclear RNA without such residues is frequently degraded within the nucleus. Most cytoplasmic messenger RNA retains this tail of poly-A sequences (although some, like histone mRNA, never seem to possess it) and one suggested function for these sequences is that they would help to control the life span of the mRNA. Perhaps one residue is docked off

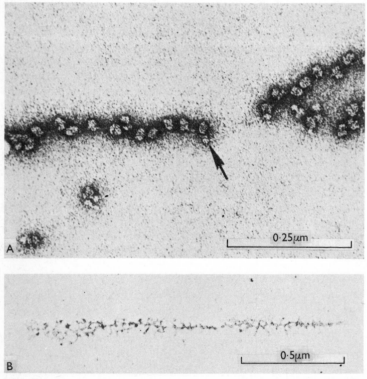

Fig. 3.19   Visualization of RNA synthesis on the DNA of *E. coli.* (A) Negatively stained (uranylacetate) portion of *E. coli* genome showing polyribosomes attached to chromosome by RNA polymerase molecule (arrow). (B) An rRNA locus showing activity of 16 S (short gradient) and 23 S (longer gradient) cistrons in log phase broth grown *E. coli.* (From Thiller, O. L. and Haenkalo, B. A. 1972. *Int. Rev. Cytol.,* **33**, 1–25. Figs. 5 and 6.)

from the tail each time the particular mRNA is transcribed, so implying a 'ticketting' function for these additional sequences.

(b) Extra-nuclear control

Those RNA molecules which are released into the cytoplasm must now be followed there in the quest for further post transcriptional controls. It must be emphasized that, in all probability, messenger RNA is not released naked into the cytoplasm. If it were, it would surely be immediately degraded by ribonuclease enzyme. In bacteria, judging from the electron micrographs of Miller (Fig. 3.19) the ribosomes 'pick-up' the messenger molecules directly from the DNA. In eukaryotes, where the nuclear membrane renders this impossible, it seems that the messenger RNA is normally conjugated with protein on its release into the cytoplasm. Scherrers' laboratory has shown particularly clearly that cytoplasmic messenger RNA is probably present in normal circumstances as a ribonucleoprotein complex (Spohr *et al.*, 1970). The protein will presumably have the dual function of protecting the RNA from breakdown and helping to recognize the appropriate ribosomal site for attachment. The life span and stability of particular mRNA molecules is itself an obvious target for control, and Tomkins *et al.* (1966) has suggested that corticosteroid hormone may affect mRNA stability.

Messenger RNA binds to the smaller of the two ribosomal subunits prior to translation, i.e. to the 30S units in bacterial ribosomes and the 40S in eukaryotic. Here it forms an initiation complex capable of binding at least one tRNA molecule prior to conjugation with the larger ribosomal subunit. Further tRNA molecules are attached only after the 30 and 50, or 40 and 60S units have combined to form the complete ribosome and following chain initiation by a special initiator methionyl tRNA. Prior to the discovery of messenger RNA, it was often supposed that ribosomes held the key to specificity for protein synthesis. Such a notion was largely abandoned in the face of mounting evidence for protein specificity being determined by the messenger RNA molecule itself. The ribosome thus becomes a non-specific assembly rig for the synthetic machinery associated with any protein type. We should now ask if this view of the ribosome is always a correct one. To be sure, the recent experiments of Gurdon's group (Lane, Marbaix and Gurdon, 1971), in which injection of globin messenger RNA into frog eggs and oocytes led to globin synthesis on a fairly large scale, leave little room for ribosomal specificity in that system. But, to be fair, there is some meagre evidence which points the other way. Bacterial ribosomes, it seems, will not handle all messenger RNA molecules with equal facility (Lodish, 1970), and some preferential binding of messenger RNA to ribosomes from the same rather than different, tissue sources, has been claimed to occur in extracts from the rat (Naora and Kodaira, 1970). So, although ribosomes will process different messenger RNA molecules, at least *in vitro*, their lack of specificity may not be complete. An idea which has often been aired is that ribosomes (or polysomes) attached to the endoplasmic reticulum might make proteins for export from the cell, whilst ribosomes (or polysomes) free in the cytoplasm might be committed to making proteins for internal use. This idea springs from the observation that cells differentiated to make a lot of protein for export are rich in endoplasmic reticulum with attached ribosomes, whilst other cells which make protein substantially for internal use are rich in free ribosomes and have a poorly developed endoplasmic reticulum. Evidence for and against such an idea can be sited, but perhaps the most compelling is that of Lisowska–Bernstein, Lamm and Wassalli (1970), who investigated the synthesis of the heavy and light immunoglobulin chains by cells of a mouse plasma-cell tumour. In this tissue, they find these secretory polypeptides are made on both free and bound ribosomes.

I have already emphasized that there is no evidence for control over the product of translation being mediated via the translation process itself. Various workers have postulated a control operating via variation in the populations of the various types of transfer RNA (see review by Sueoka and Kano–Sueoka (1970)). Such a mechanism of control seems possible. There is, for example, a report (Lee and Ingram, 1967) for a large relative difference in one of five aminoacyl-tRNAs examined, between immature red cells of the chick embryo and the reticulocytes of the adult bird. A recent paper by Manes and Sharma (1973) also demonstrates a reduced methylation of tRNA in rabbit embryos, perhaps a regulatory device.

One paragraph in this section must be devoted to the most frequently quoted example of post-transcriptional control, incidentally often cited as translational control. This is the influence of one type of globin chain on the rate of synthesis of another. During haemoglobin synthesis, the two polypeptide chains which are joined in the final tetrameric molecule are synthesized independently on different ribosomes. The rate of synthesis of the β-chains is regulated by the rate of synthesis of the α-chains, since it has been observed that a small pool of free α-chains is maintained in the cell and β-chains are only released from their polysomes when they can form dimers with the free α-chains (Fischer, Nagel and Fuhr, 1968; Maclean and Jurd (1972). Two other observations on globin chain production serve to pinpoint other aspects of control at this level. One is that during globin synthesis in erythroid cells, the growing α-chain is rapidly translated during association with comparatively few ribosomes, while the β-chain is associated with many ribosomes and is more slowly translated. This involves differences in translation rate for different polypeptide chains but is perhaps explained simply by the differing lengths of the two chains. The other observation is that α-globin can specifically regulate its own synthesis by end product inhibition, when tested in a cell free system (Blum, Malekni and Schapira, 1970). Thus a constant pool size of α-globin molecules is maintained which, by its further regulation of β-chain release, controls the overall rate of haemo-globin synthesis.

Our discussion of post transcriptional control can be concluded by stating that such control obviously operates at different levels, some nuclear and some cytoplasmic. A glance at the differentiation of *Acetabularia* (see p. 103) will assure us that such control mechanisms have important functions over comparatively long time periods and can be responsible for processes as complex as cap formation. In observing that two tissues are differentiated with respect to one another, we must not immediately conclude that the pattern of their active genes is different. All their differences could be the result of modifications of the genetic message implemented after transcription. In fact there are rather few situations in which it is possible to be sure that tissue differences in the same organism are transcriptional in origin—the dipteran polytene chromosome is an obvious case where it is possible. Perhaps the dif-ferentiated states of the reticulocyte and the erythrocyte, for example, are entirely post transcriptional in their determination.

(c) Post translational control

Modification of gene expression after translation is often ignored or forgotten as a differentiating mechanism. It should not be. There is no doubt that the pattern of enzyme activity in the cell cycle of higher cells is governed in part at this level (see reviews by Halvorson, Carter and Tauro (1971) and Tomkins (1968)). Various mechanisms may operate. Perhaps the most obvious is controlled enzymic degradation of one particular protein. The protein insulin is known to be enzymatically cleaved after its original formation to yield the active, two chain, mole-cule. In this way synthesis of one protein can also clearly govern the effective level of a second protein. It follows that the quantitative control of available protein will often be a balance between synthesis and degradation, a balance that can obviously be tipped either way accord-ing to the needs of the cell. Such a phenomenon has been recorded in a variety of cells and tissues (Tomkins *et al.*, 1972; Schimke, 1973; and Goldberg and Dice, 1974). These authors discuss three main observa-tions about cellular degradation of proteins. The first is that the rate of degradation varies widely within a cell for different proteins. That is, it is specific. Half-lives range from periods of minutes for some enzymes to 16 days for others. Secondly, the rate of breakdown of any one enzyme varies for different tissues. The isoenzyme LDH_5 has a half life of 1.6 days in rat heart, 31 days in rat muscle, and 16 days in rat liver. These figures compare with 1, 22, and 2.2 days respectively for the half life of the total soluble protein in these tissues, underlining the first point about the specificity of breakdown. Thirdly, the kinetics of pro-tein degradation so far observed are all first-order, indicating that there is no evidence for an age dependent 'ticketing' system, any molecule of a particular species, whether old or new, having an equal chance of being degraded. A tissue specific control system governing the degradation of the enzyme tryptophan pyrollase has been discovered in livers of adrenalectomized rats (Schimke, 1969). Other mechanisms which might serve for the post translational control of protein quality and quantity are the inherent instability of certain enzymes, due to loss of stabilizing ligands (see Halvorson, Carter and Tauro, 1971) or the sequestration of particular proteins by their incorporation into an insoluble matrix. This last process has been recorded in the slime mould *Dictyostelium discoideum* (Wright, Dahlberg, and Ward, 1968). Ashworth (1973) in his book *Cell Differentiation*, reviews the interesting example of the enzyme lactose synthetase; this enzyme consists of two subunits, the first of which is an active enzyme with a specific function of its own. In mammary gland cells the subunits join to yield the lactose synthetase enzyme. Even the controlled loss of enzyme through the cell membranes into the extracellular medium might be used as a regulatory mechanism governing effective enzyme levels. By such a process, cells could become widely differentiated from one another by a fairly simple change in the properties of the cell surface, an obvious target for environmental influence via hormones or other surface active molecules.

This list of possible post-translational control mechanisms is certainly not exhaustive, and no doubt in many situations a number of the mechanisms cited might act co-ordinately. In addition where proteins are modified following their release from the ribosome, the mechanism inducing the modification remains obscure. For example, the enzyme glutamate dehydrogenase can be recovered from the cytoplasm in a wide range of polymer sizes, and such molecules as nucleotides and steroids are known to have marked effects on polymer size. Similarly isoenzymes of lactate dehydrogenase, consisting in some cases of monosomes and in others of dimers of the same polypeptide, occur in varying ratios depending on the immediate ionic environment of the cell (Shaw, 1969).

4

The cytoplasm in differentiation

Having reviewed differentiation as a function of nuclear gene expression in Chapters 2 and 3, we will now examine in detail the important role of the cytoplasm.

1926 was an important year in the history of cytology, for it was then that the Feulgen stain for DNA was discovered. This, for the first time, demonstrated that the nucleus contained DNA. Indeed it appeared to contain all the DNA of the cell. Also in that same year, the geneticist, T. H. Morgan, declared that 'the cytoplasm may be ignored genetically' (Morgan, 1926). We now know that both the colorimetric observation and the statement of approach are seriously misleading: DNA is present in the cytoplasm as well as the nucleus of eukaryotic cells and in many instances the cytoplasm plays an important role in hereditary determination. Moreover, even if the significance of the cytoplasm in heredity is second to that of the nucleus, this is far from the case in studies on differentiation. Modification of nuclear activity by the cytoplasm is every bit as important as the modification of cytoplasmic form and function by the nucleus. There are famous experiments illustrating both processes in differentiation.

The first involves nuclear transplantation between two species of *Acetabularia*, (see p. 61), a large unicellular marine alga consisting of a rhizoid or hold fast, a stalk, and an umbrella-shaped cap. The form of the cap differs between the two species *Acetabularia mediterranea* and *crenulata* (Fig. 4.1). Nuclear transplantation is relatively easy in these cells, since the very large nucleus is located in the basal rhizoid. The classic experiment, first carried out by Hämmerling (1934) involves removing the caps from individuals of the two species and reciprocally exchanging their nuclei by grafting the stem of one on to the rhizoid of the other.

In such graft hybrids, initial cap formation was of a form intermediate between the two species, but when the hybrid caps were removed and the stems back-grafted on to their correct rhizoids the cap which now formed was typical of the species donating the rhizoid. The results of this experiment are usually taken to imply that the form of the cap is determined solely by the nucleus. On the other hand, cytoplasmic determination of cellular morphology is well illustrated by the

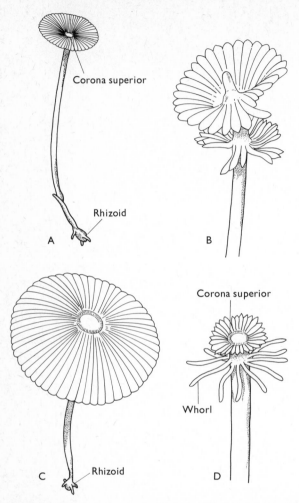

Fig. 4.1 Regeneration in *Acetabularia*. (A) *Acetabularia mediterranea* regenerated from a rhizoid. (B) *A. crenulata* with whorl formation beginning. (C) Regenerated *A. mediterranea* with very short stalk. (D) Young cap of *A. crenulata* with pointed rays and remains of old whorl. (After Berrill, N. J. 1961. *Growth, Development and Pattern*. W. H. Freeman and Co. San Francisco.)

transplantation experiments of Gurdon. When nuclei from intestinal epithelial cells of larval *Xenopus laevis* were transplanted into enucleated *Xenopus* eggs (Fig. 4.2) some of these eggs proceeded to divide and produce normal larval and adult *Xenopus*. It appears undeniable that the egg cytoplasm has been responsible for ensuring that the nucleus alters its form and function from that normal for an intestinal epithelial cell (Gurdon and Uehlinger, 1966).

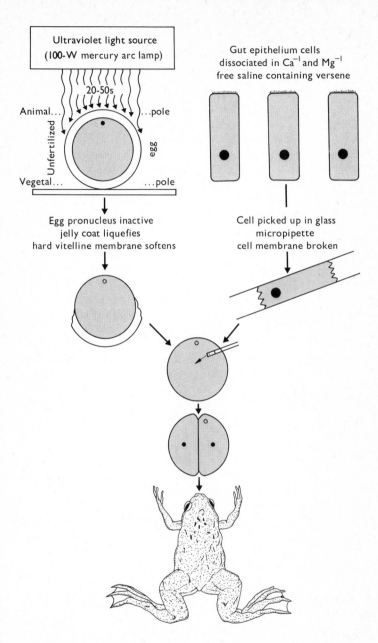

Fig. 4.2 Summary of the techniques used by Gurdon and Uehlinger in preparing nuclear-transplant frogs of *Xenopus laevis*. (After Whittaker, J. R. *Cellular Differentiation*. 1968. Dickenson Publishing Co. Inc.)

Cytoplasmic activity in differentiation can be viewed at three levels. Firstly, the modification of nuclear genes by cytoplasmic molecules, secondly the role of pre-existing cytoplasmic structures in determining the form of new structures, and thirdly the role of cytoplasmic DNA.

Our discussion of activity at the first level will be rather brief in this chapter, since we have already reviewed this area in our study of nuclear gene activity in previous chapters (see pp. 34 and 63). Of course it is not always clear in any particular biological situation, which level of cytoplasmic activity is involved, and in many cases more than one may apply. For example, the important contribution made by the cytoplasm to the separate differentiation of sections of the highly mosaic egg may involve activity at all three levels. Many examples of so-called maternal inheritance, in which the egg cytoplasm determines characters revealed in the progeny, are not sufficiently understood for us to be able to determine the level or manner of the cytoplasmically transmitted gene activator, the precise cytoplasmic template or the cytoplasmic gene system involved. Such examples include the remarkable one-way sterility found in reciprocal crosses between geographical populations of the mosquito *Culex pipiens*, and also of the willow herb, *Epilobium hirsutum*. The same statement holds for the cytoplasmic inheritance of the 'mi' variant in the fungus *Neurospora crassa*. These, and other, examples are discussed more fully by Jinks (1964). Again, the nuclear gene controlling the left or right handed direction of the spiral cleavage of the embryo water snail *Limnea*, and ultimately also the left or right handed spiral of the shell, has been found to be active via the egg and not the sperm (Fig. 4.3, and see Boycott and Diver, 1923).

Yet another situation in which a relatively obscure cytoplasmic factor plays a crucial role in differentiation is in the somatic and germ cell lines of the gall midges *Cecidomyiidae*, in which a curious diminution of chromosome number occurs in the somatic cell line (see p. 65). It appears that the maintenance of the full chromosome complement in the germ line depends upon an aggregation of cytoplasmic material which can be visualized microscopically. The nucleus in closest proximity to this cytoplasmic aggregate is protected from chromosome diminution, and a shift of the aggregate by centrifugation involves altered activity (see discussion in Jinks, 1964). Although the constituents of this active cytoplasmic aggregate have not been fully analysed, it is known to be rich in RNA (Ullmann, 1965).

4.1 Modification of nuclear gene activity by cytoplasmic molecules

Since nuclear proteins, both acidic and basic, are synthesized in the cytoplasm of the eukaryotic cell, it follows that control of gene expression by these proteins may ultimately be determined in the cytoplasm. Appreciation of this fact has probably contributed to the idea of

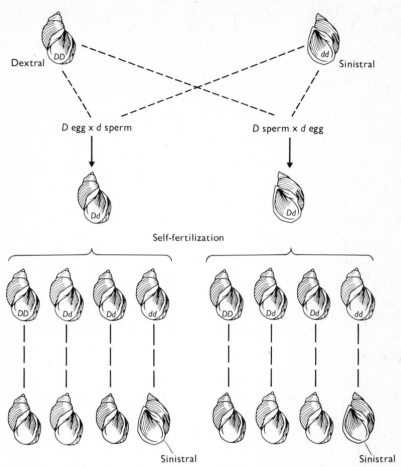

Fig. 4.3 Inheritance of coiling in *Limnaea peregra*. (After Strickberger, M. W. 1971. *Genetics.* Collier-Macmillan Ltd., London.)

altered gene expression being effected at or about the time of mitosis, when the absence of the nuclear membrane would permit easy access of cytoplasmic proteins to the DNA. Many cytoplasmic effects on nuclei can be interpreted as the effects of cytoplasmic-derived protein within the nucleus and, if current ideas on the matter are correct, the acid protein activates genes in transplanted nuclei partly by altering the histone binding at particular loci. If nuclei are transplanted into cells containing histone tagged with radioactive iodine, the histone is found to enter the new nucleus and accumulate in it (Gurdon and Woodland, 1970). Similar experiments were carried out by Arms in 1968 by the injection of brain nuclei into unfertilized eggs previously labelled with tritiated leucine (Fig. 4.4). Labelled protein was found to have passed from egg cytoplasm into the brain nucleus. Of course gene activation or repression by RNA may involve RNA molecules which have never left the nucleus, but in both induction and repression of bacterial operons the apo-regulator molecule appears to be a protein, and so it may be that in eukaryotes gene activation or repression via nuclear RNA will still require the participation of a protein supplied by the cytoplasm.

The plasma membrane obviously fulfils a highly important function in determining which molecules enter or leave the cell; the nuclear membrane fulfils a similar function between the cytoplasm and nuclear contents. Insertion of a nucleus into foreign cytoplasm could conceivably deprive some genes of the raw materials needed for their expression. Moreover, the involvement of the cell membrane in recognition, attachment and activity of the polypeptide hormones, and in attachment and transport of steroid hormones into the cell, is obviously crucial. The differential response of tissues to the same hormone must often, and probably most frequently does, involve purely cytoplasmic factors. This question is discussed at greater length in Chapter 5.

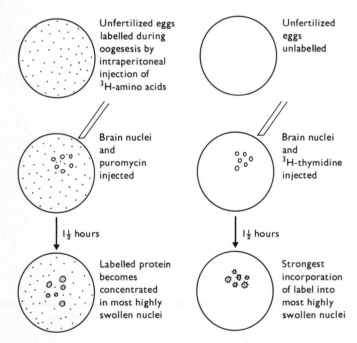

Fig. 4.4   Injection of brain nuclei into unfertilized eggs, followed by assay for DNA synthesis. Puromycin effectively suppresses protein synthesis. (After Gurdon, J. B. and Woodland, H. R. 1970. *Current Topics in Devl. Biol.*, **5**, 57. Academic Press Inc.)

4.2 Pre-existing cytoplasmic structures which are necessary for the synthesis of new structures

The degree to which cytoplasmic structures act as templates is not known with any certainty and experiments to investigate the problem are difficult to design. Presumably if isolated nuclei could produce cytoplasmic structures when suspended in a culture medium, we might legitimately conclude that no pre-existing cytoplasmic template was required for complete assembly, though such an experiment would yield more problems than answers. Certainly such structures as endoplasmic reticulum and nuclear and plasma membranes, may well require existing membranes for their correct assembly, although egg and embryonic cells often have very little endoplasmic reticulum.

In a few examples, however, fortuitously favourable conditions point to a clear need for pre-existing templates for the elaboration or organization of further structure. One such instance is the cell wall of bacteria. Many bacteria, when treated with penicillin, lose their cell walls. This is because penicillin acts by inhibiting the terminal cross-linking step in the synthesis of murein, an important constituent of the bacterial cell wall (Fig. 4.5). If certain types of bacteria are treated with high concentrations of penicillin in a medium of sufficiently high osmotic strength, then the bacterial cells, devoid of cell walls, round up but continue to survive, and are then known as protoplasts. Naked bacterial protoplasts are often referred to as L-forms, following their discovery in the Lister Institute by Klieneberger-Nobel in 1935. Now, if the exposure

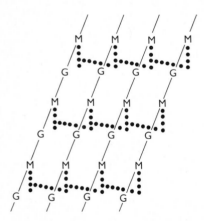

Fig. 4.5 Cell wall assembly in bacteria. A schematic representation of the murein of *Staphylococcus aureus*. In this species, the tetrapeptide chains (–L–ala–D–glu–L–lys–D–ala), symbolized by vertical rows of four dots attached to muramic acid residues (M), are almost completely cross-linked to one another by pentaglycine bridges, symbolized by horizontal rows of five dots. (After Ghysen, J. M. 1968. *Bacteriol. Revs.*, **32**, 425–64. Fig. 1.)

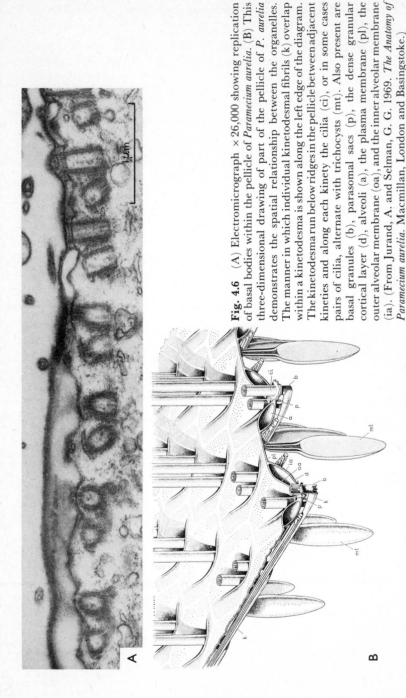

Fig. 4.6 (A) Electromicrograph ×26,000 showing replication of basal bodies within the pellicle of *Paramecium aurelia*. (B) This three-dimensional drawing of part of the pellicle of *P. aurelia* demonstrates the spatial relationship between the organelles. The manner in which individual kinetodesmal fibrils (k) overlap within a kinetodesma is shown along the left edge of the diagram. The kinetodesma run below ridges in the pellicle between adjacent kineties and along each kinety the cilia (ci), or in some cases pairs of cilia, alternate with trichocysts (mt). Also present are basal granules (b), parasomal sacs (p), the dense granular cortical layer (d), alveoli (a), the plasma membrane (pl), the outer alveolar membrane (oa), and the inner alveolar membrane (ia). (From Jurand, A. and Selman, G. G. 1969. *The Anatomy of Paramecium aurelia.* Macmillan, London and Basingstoke.)

of the normal cells to penicillin is brief, most of the cells continue to make new cell wall and repair the penicillin damage to the cell. However, if penicillin exposure is sufficiently prolonged, permanent L-forms are established and resynthesis of cell wall does not occur (Stanier, Doudoroff and Adelberg, 1971). This permanent lack of the wall seems to be explained by the complete destruction of the existing wall by the penicillin. The enzymes which are responsible for laying down the murein layer apparently require an existing murein template on which to build additional material. Loss of template means permanent loss of the structure.

Such a phenomenon is not confined to bacterial cells since a characteristic feature of all polysaccharide synthesis is the requirement for a short segment to act as primer. Primer requirement has been studied in some detail in glycogen synthesis, and it has been found that effective primer action depends on the existence of a structure of at least four sugar units. The characteristic branched form of the glycogen molecule is achieved by special branching enzymes which detach small fragments from the end of the polysaccharide chain and reinsert them elsewhere in the chain. We can surely conclude that if, during development, any cell is left without glycogen, even the activation of all nuclear genes will not achieve further glycogen synthesis. The loss of the template would mean permanent cytoplasmic differentiation.

Two other organelles found in eukaryotic cells may fall within this section. They are centrioles, and the basal bodies or kinetosomes of cilia. Many unanswered questions remain in connection with these structures. First, the extent to which the existence of a centriole or basal body is *necessary* for the synthesis of an additional one is debatable, since there is evidence which seems to show that centrioles, which are absent from unfertilized sea urchin eggs and are normally donated by the sperm at fertilization, will arise *de novo* if the eggs are artificially activated (Pitelka, 1969). But there seems little doubt that the existing organelles are often, if not always, involved in the assembly of new ones (Fig. 4.6). The second problem centres on the question of whether or not they possess their own DNA. Such a claim has been advanced on a number of occasions, and is not easy to disprove entirely. However, the weight of evidence now rests rather heavily against the idea, at least in basal bodies (Younger, Banerjee *et al.*, 1972 and Flavell and Jones, 1971). So we will come down tentatively on the side of centrioles and basal bodies lacking DNA, but will also assume that their involvement in their own replication is often crucial.

Centrioles and basal bodies belong together in both structure and function, since they both produce cellular fibrous structures—asters in the case of centrioles (and perhaps also the spindle fibres) and ciliary fibres in the case of basal bodies. Indeed there is a continuity of microtubules running through the centriole and aster apparatus, and the basal body and cilium which it supports. Moreover, at least in sperm, the centriole acts like a basal body in the production of the sperm tail, an organelle which is not unlike a cell cilium in its fine structure.

For our purpose the main point of interest in centrioles and basal bodies lies in their role in the replication of new structures. The most convincing evidence that these structures are involved in the heritable properties of the cell comes from research on ciliated protozoa. In these organisms the basal bodies are arranged in long parallel rows called kineties and individual basal bodies certainly seem to be implicated in the generation of new daughter basal bodies around them (Fig. 4.6). The ciliated protozoans, *Paramecium* and *Tetrahymena*, have variable numbers of kineties (rows of cilia) in their pellicle, and these appear to be passed on to the progeny in a consistent manner (Fig. 4.7). If *Paramecium* or *Tetrahymena* with differing numbers of kineties are allowed to conjugate, the ex-conjugants retain their original pattern of kineties and continue to faithfully pass it on to their daughter cells (Sonneborn, 1963; and Nanney, 1966). Conjugation in these ciliates involves exchange of nuclear genes and also some cytoplasm, and so it seems inescapable that the cortex controls its own kinety pattern independently and that new basal bodies arise from pre-existing ones without interference from nuclear genes. Of course the precursors used for assembly are no doubt the products of information stored in nuclear genes, but the pattern of assembly is determined by the existing cytoplasmic structures.

We can conclude that some cytoplasmic structures not only help in the assembly of further structures but that, during cell division, unequal assortment of some of these structures, and especially their absence from a daughter cell, would confer a permanent differentiation pattern in all the future cell generations. Cells without certain polysaccharide molecules may be absolutely incapable of producing cells with them, despite the presence of all necessary nuclear genes. So too, cells with a set number of cilia will always produce daughter cells with that number, whatever the genetic information in the genome.

4.3 Heterokaryons and nuclear transplants

We have already cited several examples of nuclear transplantation into new cytoplasm as evidence of cytoplasmic influence on differentiation. These include *Acetabularia* nuclear transplants; transfer of nuclei from *Xenopus* intestinal, and other, cells into enucleated eggs, and injection of brain nuclei into unfertilized eggs. Artificial heterokaryons made between chicken erythrocyte nuclei and a variety of other cells have also been briefly discussed in Chapter 2.

Since the exposure of the nucleus to new cytoplasm is potentially so informative about the role of cytoplasm in differentiation, we will now review more fully the conclusions which such experiments permit.

A word of caution is appropriate here about the interpretation of

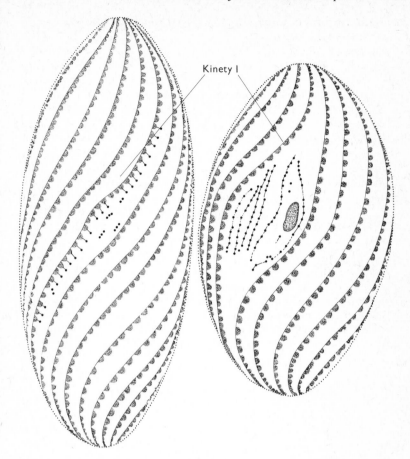

Fig. 4.7 Kinety replication in a ciliate. Formation of oral ciliature in *Chromo-dina* by division of kinetosomes of kinety 1. (After Berrill, N. J. 1961. *Growth, Development and Pattern.* W. H. Freeman and Co. San Francisco.)

evidence in nuclear transfer experiments where the resulting cell is multinucleate. In autoradiographic experiments with transplanted nuclei of *Amoeba*, Byers, Platt and Goldstein (1963) obtained evidence for migration of protein molecules from one nucleus to another within the same cell. The function or nature of these proteins remains unclear, but it is obvious that allegedly cytoplasmic influences on synthetic syn-chrony or other effects within nuclei in multinucleate cells must be viewed with caution. In some cases, the cytoplasm might act only as a medium through which migrating molecules could pass from one nucleus to another.

(a) The cytoplasm affects nuclear size, in terms of both volume and mass

The ratio of nuclear volume to cytoplasmic volume is something of a chicken and egg situation, since large nuclei are normally situated in large volumes of cytoplasm and *vice versa*. Moreover, large nuclei are most frequently those containing very high amounts of DNA. For example, the amphibian *Amphiuma*, with a 2C value of 168 and the plant *Lilium longiflorum* with a 2C value of 134, have very large cells. Likewise polyploid cells are generally larger in both nuclear and cytoplasmic volume than their diploid counterparts. However, transfer of a nucleus to a new and different cytoplasmic environment shows that nuclear size can be altered. An initial swelling of the nucleus is often reported, particularly where nuclei are obtained from relatively inactive cells such as chicken erythrocytes and incorporated into macrophages or HeLa cells (Harris, 1970). Moreover, in such a transplant, the dry mass of the erythrocyte nucleus is found to increase greatly, presumably due to the entry of cytoplasmic molecules (Bolund *et al.*, 1969). Although some nuclei fail to swell in new cytoplasm, actual nuclear shrinkage has not been recorded.

The only example known to the author in which alteration in cytoplasmic volume has been accomplished experimentally without actual transplantation or cell fusion is in regard to the polar lobe of annelid and mollusc embryos. Many of these embryos possess, during early cleavage, a large lobe of cytoplasm asymmetric to the main axis of the egg embryo (Fig. 4.8). The polar lobe has been intensively studied in the marine snail *Illyanassa*, and experiments undertaken in which polar lobes were removed. The most cogent evidence here is that of Davidson *et al.* (1965) who found greatly diminished levels of RNA synthesis in the nuclei of lobeless embryos. However, parameters as general as those of nuclear or cytoplasmic volume are obviously tied up with many other critical factors in the study of nucleo-cytoplasmic relations, and we must be content to conclude simply that nuclear and cytoplasmic volumes are of some importance in cell function, and are frequently interrelated.

(b) The cytoplasm affects the timing of DNA synthesis

During the cell cycle of eukaryotic cells, DNA synthesis occurs in a rather confined period, referred to as the S period. The timing of the S period appears to be very greatly affected by the cytoplasm. Several lines of evidence lead to this conclusion. Firstly, in a variety of naturally multinucleate cells, such as Protozoa, Fungi, plant endosperm, animal germ cells, and some tissue culture cells, DNA synthesis occurs synchronously in all nuclei in the same cytoplasm (Johnson and Rao, 1971). A similar situation obtains in cells which are made artificially multinucleate by nuclear transfer. These include *Amoeba*, *Stentor*, slime moulds, and a great variety of heterokaryons made by virus-induced cell fusion.

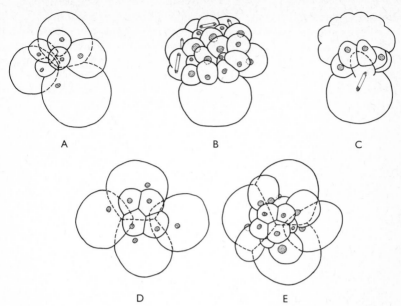

Fig. 4.8 Cleavage of the snail *Ilyanassa*, showing development with and without the polar lobe. (After Davidson, E. H. 1969. *Gene Activity in Early Development.* Academic Press, London, New York.) (From camera lucida drawings of stained whole mount preparations) × 326. (A)–(C) normal cleavage. (D)–(E) Cleavage after removal of the polar lobe at the trefoil stage.

It is particularly striking to find that even when heterokaryons are formed with nuclei taken from amoebae of different species, DNA synthesis and mitosis become synchronized (Jeon and Lorch, 1969). In general, it may be said that DNA synthesis takes place synchronously in all G_1 or S nuclei within the same cytoplasm. This is most clearly demonstrated by nuclear transplants in *Stentor* (De Terra, 1970) and by fusion of tissue culture cells (Rao and Johnson, 1970). If, however, G2 nuclei are combined with S or G_1 nuclei, DNA synthesis, as shown by both cell fusion experiments (Johnson and Rao, 1970) and nuclear transplant experiments (Ord, 1971), does not become synchronized until mitosis has intervened.

(c) The cytoplasm affects chromosome condensation and mitosis

The interval between DNA synthesis (S period) and mitosis is termed G_2. In some cells G_2 is very short; but in other cells it is lengthy, and this enables us to study these two stages in the cycle separately. Many, though not all, of the artificial heterokaryons and multinucleate cells in which synchrony of DNA synthesis has been observed, also display synchrony of chromosome condensation, a normal prerequisite for mitosis or meiosis, or both. It is also appropriate to remark here that one of the almost universal components of mitosis is the breakdown of the nuclear membrane, and that this event exposes the chromosomes to cytoplasmic constituents from which they had been previously separated. Mitosis must surely be a time when cytoplasmic influences on the nuclear material are maximized, although the effects would be seen only post mitotically.

(d) The cytoplasm affects nuclear RNA synthesis

To say that the cytoplasm affects nuclear RNA synthesis is simply to state the obvious, but more detailed information derived from nuclear transplants is worth discussing. Gurdon's laboratory at Oxford has been particularly active in seeking this sort of information (Gurdon and Woodland, 1968; and Gurdon, 1974), and they have carried out complementary experiments. On the one hand, in order to demonstrate cytoplasmic induction of RNA synthesis, *Xenopus* mid-blastula nuclei, normally making very little RNA, were injected into oocytes, in which RNA synthesis is normally very vigorous. Within three hours of injection, the previously low level of RNA synthesis had greatly increased in these transplanted blastula nuclei, apparently under the influence of the oocyte cytoplasm. The converse experiment consisted in removing nuclei from gastrula or neurula endoderm, again in *Xenopus*, and injecting these nuclei into *Xenopus* eggs. Whereas the egg is very inactive in RNA synthesis, the gastrula and neurula are highly active, particularly in ribosomal RNA and transfer RNA production respectively. In these experiments, Gurdon has demonstrated considerable inhibition of both ribosomal and transfer RNA manufacture after exposure of the nuclei to the egg cytoplasm.

Another experimental approach can appropriately be discussed here although it does not involve transplantation of individual nuclei. It consists in making bulk preparation of isolated nuclei and exposing them, under suitable controlled conditions, to cytoplasmic extracts prepared by ultracentrifugation. The activity of the nuclei, with or without cytoplasm, is then measured by the incorporation of labelled uridine triphosphate (UTP) into RNA. Since such a method is likely to yield information only about general increase or decrease in incorporation, most laboratories using this approach have employed nuclear preparations from chicken erythrocytes (Thompson and McCarthy, 1967; Leake, Trench and Barry, 1972), working on the assumption that such nuclei do not normally synthesize any RNA. In fact this assumption is not valid (Madgwick, Maclean and Baynes, 1972) and so the evidence for this 'reactivation' by exposure to cytoplasmic extracts should be interpreted as, at best, enhancement. Probably a similar criticism can be levelled at the claims for reactivation of RNA synthesis in the chicken erythrocyte nucleus within heterokaryons (Johnson and Harris, 1969). In my own laboratory we began 'reactivating' chicken erythrocyte nuclei by exposure to cytoplasm and were surprised at the high level of incorporation in our control nuclear preparations. On further investigation we find that the intact chicken erythrocyte does make RNA, and the apparent inactivity of its nucleus after isolation must be interpreted as resulting from the manner of its isolation. We are now proceeding to use *Xenopus* erythrocyte nuclei instead, since their intrinsic level of RNA synthesis is much lower than that of the chicken (Maclean, Hilder and Baynes, 1973). With this system we do find that, if the nuclei are exposed to liver cytoplasm, then washed free and provided with tritiated UTP in a suitable medium, their RNA synthetic activity greatly exceeds the minimal incorporation detectable in the unexposed nuclei (Fig. 4.9; see also Hilder and Maclean, 1974). Although it is premature to read much into these results, they do reinforce the idea that cytoplasm can affect the quantitative and probably the qualitative control of RNA synthesis by the nucleus.

That cytoplasm controls the quantitative synthesis of nuclear RNA is further confirmed by nuclear transfer experiments in amoebae. When heterophasic homokaryons (cells artificially formed by nuclear transfer so as to include multiple nuclei derived from cells in different stages of the cell cycle) are exposed to tritiated uridine all nuclei sharing the same cytoplasm incorporate the same amount of H^3UdR regardless of their cell cycle phase (Ord, 1973).

(e) The cytoplasm, in some cases, diverts the direction of differentiation

Many of the properties already attributed to the cytoplasm, in its influence on the nucleus, are likely to involve the direction of differentiation. Certain results which suggest such an influence being exerted by the cytoplasm, are not immediately attributable to altered cell cycle metabolism or nucleic acid synthesis. Again, the method is not strictly that of nuclear transplantation, but rather of spatial manipulation of nuclear material within the cell. Originally carried out by Carlson in 1952 the experiment consisted of rotating the mitotic spindle of a dividing grasshopper neuroblast cell with a microneedle. The neuroblast normally divides unequally, one daughter cell persisting as a neuroblast and capable of further division, while the other differentiates into a ganglion cell. Some anticipation of this differentiation is detectable in the dividing neuroblast, since the cytoplasm at the potential ganglion cell end is visibly different from the cytoplasm at the other pole. By rotating the spindle at anaphase, the chromosomes destined to migrate to the ganglion cell pole were redirected to the other pole, and *vice versa*. Carlson found that the micro-manipulation of the chromosomes made no difference to the direction of differentiation, the ganglion cell pole cytoplasm continuing to produce a ganglion cell with either set of chromosomes. This experiment serves to confirm that the determination for

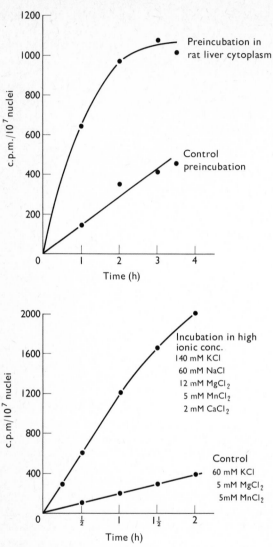

Fig. 4.9   Reactivation of *Xenopus* erythrocyte nuclei by exposure to ionic solutions and cytoplasmic extracts.

a particular developmental pathway is often engineered by first committing the cytoplasm rather than the nucleus. No doubt the determination of direction was originally invoked in the cytoplasm partly at the instigation of the nuclear genes, but, at least at a temporal level, there is a period in differentiation during which the cytoplasm directs the nucleus and hence the developmental destiny of the cell.

4.4 Cytoplasmic DNA

Two very well authenticated sites for cytoplasmic DNA are now known —mitchondria and chloroplasts. Two other more doubtful cases are the informosomes or i-somes proposed by Eugene Bell and the report from Brachet's laboratory of DNA associated with yolk platelets in the egg (Brachet and Ficq, 1965). Bell's proposal rests on his ability to recover rather low molecular weight thymidine-labelled molecules from the cytoplasm of embryonic chick cells. He argues that these represent an amplification system in which copies of genes crucial for certain developmental stages are released into the cytoplasm for rapid transcription (Bell, 1971). The reports of DNA associated with yolk platelets are most easily explained by assuming the DNA to be a residue of the mitochondrial DNA known to be present in great excess in these oocytes. Mitochondria are a probable source of much of the yolk protein. We must await further evidence before seriously considering either i-somes or yolk platelet-associated DNA as serious contenders for cytoplasmic DNA constituents.

In addition to these examples, there are other transient cytoplasmic DNA factors such as kappa particles in *Paramecium* and sigma factor in *Drosophila*. Since mitochondria are in many ways analogous to symbiotic bacteria, the bacteria-like infective factors which contain DNA, such as kappa, will be considered in our discussion of mitochondria in this chapter, while the virus or virus-like cytoplasmic DNA factors will be discussed in Chapter 6 on episomes and viruses.

Mitochondria and chloroplasts undoubtedly contain DNA (see Fig. 4.10) and, although these organelles do have some features in common, we will discuss them separately.

(a) Mitochondria

Mitochondria (Fig. 4.11) play an important role in differentiation, controlling as they do many of the important enzyme pathways of the cell. Table 4.1 provides a brief catalogue of the characteristics of mitochondria and, on the basis of this evidence, probably most biologists now believe mitochondria to be the descendants of symbiotic bacteria within eukaryotic cells. The pros and cons of the argument need not concern us here, but it is worth emphasizing the importance of these highly autonomous organelles for cellular differentiation. An appreciation of their autonomy implies that a differentiating cell must manipulate the form and activity of its mitochondria in tune with its own development. The degree to which the observed changes in cellular mitochondria are a direct result of the differential action of nuclear genes, or simply the relatively passive adaptation of the mitochondrion to a changing cytoplasmic environment, is not known. But an equally important consequence of mitochondrial autonomy is that, in at least some cases, cellular differentiation could be purely a result of mitochondrial

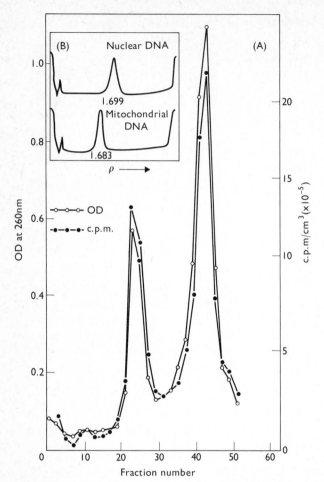

Fig. 4.10 Separation of nuclear and mitochondrial DNA from normal yeast cells.

(A) Separation of nuclear amd mitochondrial DNA by density gradient centrifugation. DNA from *Saccharomyces cerevisiae* (350 μg) was mixed with $HgCl_2$ (145 nmmol) and centrifuged to equilibrium in Cs_2SO_4 (1554 g/cm^3) at 68 000 g for 48 hours. After centrifugation, 10-drop fractions were collected and diluted with 0.5 cm^3 of water. Optical density at 260 nm and radioactivity of each sample was determined.

(B) Densitometer tracings of ultraviolet photographs of purified nuclear and mitochondrial DNA of *S. cerevisiae*. After centrifugation to equilibrium in CsCl (1705 g/cm^3, pH 8.5) at 44 770 rpm for 17 hours in a Beckman model E analytical centrifuge, ultraviolet photographs were taken and scanned on a Joyce-Loebl microdensitometer. (After Padilla, G. M., Whitson, G. L. and Cameron, I. L. 1969. *The Cell Cycle—Gene-Enzyme Interactions*, Academic Press, London and New York.)

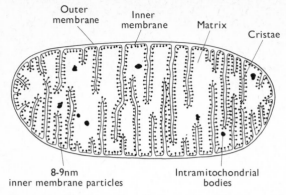

Fig. 4.11 Diagram of a typical mitochondrion as seen from an electron micrograph. (After Loewy, A. G. and Siekevitz, P. 1969. *Cell Structure and Function*. Holt, Rinehart and Winston, London and New York.)

Table 4.1 Properties of mitochondria relevant to their possible affinity with bacteria

1   Mitochondria contain DNA with a base composition different from that of the nuclear DNA, enabling its isolation as a distinctive satellite band in centrifugation through caesium chloride density gradients (Fig. 4.10). In animal cells it is normally circular and about 5 μm long—some 15 000 base pairs, but in higher plants and *Neurospora* the mitochondrial DNA has a much higher molecular weight and has not yet been proved to be circular. *Tetrahymena*, too, has a rather larger mitochondrial DNA molecule than other animal cells. The mitochondrial DNA in animal cells is only about 0.01 of that in the bacterium *E. coli*, but plant cell mitochondria more nearly approach *E. coli* in the length of their DNA.

2   Mitochondrial DNA replicates out of synchrony with the normal S period of nuclear DNA synthesis.

3   The DNA of mitochondria, like that of prokaryotes, is unconjugated with histone, while that of almost all eukaryotic cell nuclei is associated with histone.

4   Mitochondrial ribosomes have an S value nearer to that of the bacterial than the normal eukaryotic ribosome, and the ribosomal RNA of mitochondria is coded on the mitochondrial DNA.

5   Mitochondrial transfer RNA molecules are distinct from those in the rest of the cell, and at least some of them are coded on the mitochondrial DNA. N-formylmethionyl transfer RNA occurs only in bacteria and eukaryotic mitochondria.

6   About 80% of mitochondrial protein is synthesized on cytoplasmic ribosomes, the remainder on the mitochondrial ribosomes.

7   Chloramphenicol inhibits amino acid incorporation in bacteria and mitochondria but not in the cytoplasm of the eukaryotic cell.

Table 4.1 (*contd.*)

---

8   Mitochondria possess a double limiting membrane. The outer one is sometimes continuous with that of the endoplasmic reticulum, and is morphologically different from the inner membrane. Mitochondrial inner membranes are often convoluted into folds or cristae, and contain the major components of the respiratory chain enzyme system. Both of these features apply also to the bacterial cell membranes (Fig. 4.11).

9   Bacteria and mitochondria are of rather similar size ranges, about 1–5 μm.

10   The morphology of mitochondria is variable, both between tissues of the same organism and between, say, an amoeboid and a mammalian cell. Put in more relevant terms, mitochondria differentiate both during the evolution of organisms and in the development of different tissues within one organism.

11   The manner of initiation of protein synthesis on the ribosome is similar in bacteria and mitochondria.

12   The nuclear genome of *Xenopus* does not possess a copy of the mitochondrial DNA (Dawid, 1972).

13   Mitochondria divide within the cytoplasm by a fission process.

14   The cytochrome enzymes are coded on the nuclear genome but only become metabolically active in an intact mitochondrion. (Woodward, Edwards and Flavell, 1970). Petite mutant yeast lack active cytochrome enzymes (see p. 124).

---

differentiation. Either differing populations of mitochondria may exist in all organisms and these may be allotted unequally to the differentiating tissue cells during development; or a progressive differentiation of mitochondria may occur during development, and this in turn would induce alteration in the cells within which the mitochondria reside. A third possibility is that some cellular differentiation may be the consequence of the unequal segregation of mitochondria at cell division, cells having less mitochondria becoming distinct from those having more.

It will be apparent from the preceding paragraph that almost all we have to say in this book about cellular differentiation could be applied to the mitochondria alone, and therefore that we, as scientific observers, are suddenly confronted with a Russian doll situation of a differentiating cell system within a differentiating cell system.

Most of what has so far been said about mitochondria in differentiation is speculative. What are the facts? In what ways do mitochondria and their DNA participate in or determine cellular differentiation?

Firstly, there are a few examples of a very unequal segregation of mitochondria during early cleavage. In the oligochaete worm, *Tubifex*, at the first cleavage of the fertilized egg, most of the mitochondria go into one of the daughter cells, and all the embryonic mesoderm is ultimately derived from that cell. A similar unequal distribution of mitochondria occurs in the first cleavage of the ctenophore *Beroë* and, in this case, the mitochondria-rich cells are those finally responsible for cilia production. The correlation of a large mitochondrial complement in a cell and a particular direction of differentiation has also been demonstrated experimentally. If the zygotes of ascidians are gently centrifuged, the mitochondria are concentrated at one end of the cell and are then donated mainly to one daughter cell at cleavage. The cell receiving the large assignment of mitochondria determines the development of muscle later in embryological development. We can conclude that an unequal distribution of mitochondria may frequently be used as a differentiating device, although continuing to recognize that it can equally well be a result rather than a cause of differentiation.

Secondly, the appearance of very small colonies in agar plates of the budding yeast *Saccharomyces cervisiae* provides interesting information relating to the degree of mitochondrial autonomy. These small colonies are termed petites, and they are to be found occasionally if quite normal cultures of yeast are plated out. However, if yeast cells are pre-treated with, for example, acriflavine or UV-irradiation, the percentage of petites is very greatly increased. Two separate classes of petite colony have now been defined (Fig. 4.12). The one class, termed segregational petites, are clearly the result of nuclear gene mutation, as evidenced by their breeding behaviour. The cells of petite colonies often lack many cytochrome enzymes and these are known to be coded by nuclear DNA. The other, termed non-segregational petites, have suffered only a cytoplasmic effect, since interbreeding with wild-type cells leads to completely wild-type progeny. Many petites totally lack mitochondral DNA (Fig. 4.13). In non-segregational petites, therefore, wild-type cells have presumably donated functional mitochondria to all daughter cells. Other interesting curiosities of the biology of yeast mitochondria are (1) the appearance of erythromycin-resistant cells (erythromycin is lethal to mitochondria, as it is to bacteria) which are able to endow sensitive cells with resistance to the antibiotic cytoplasmically, no doubt by passing on resistant mitochondria, and (2) the observation that chloramphenicol-inhibited yeast cells, which have 'normal' but nonfunctional mitochondria, grow much faster than petite cells which lack functional mitochondria. This seems to infer an active role for the mitochondrial DNA even when mitochondrial protein synthesis is entirely suppressed.

Mutants of the fungus *Neurospora*, termed 'poky', resemble the petite mutant in yeast in many ways, while clones of *Paramecium aurelia* which are resistant to erythromycin have been reared by Beale (1969). Such cells donate their erythromycin resistance to their partner cells at conjugation. These observations show that, as in yeast cells, mitochondrial properties are passed on via the cytoplasm, and are at least partly independent of nuclear gene control.

A third study relating mitochondria and differentiation is that involving interspecific crosses between *Xenopus laevis* and *mulleri* (Table 4.2). These crosses have been extensively studied in the laboratory of Igor

124

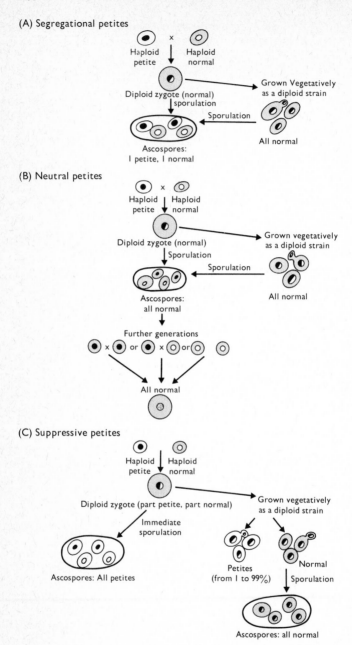

Fig. 4.12 Results of crosses between different kinds of 'petites' and normal strains of yeast. (After Wilkie, D. 1964. *The Cytoplasm in Heredity*. Methuen, London.)

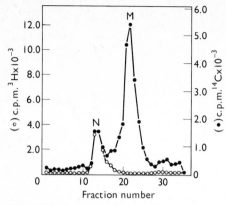

Fig. 4.13 Double label caesium chloride whole cell DNA of two strains of budding yeast, one without mitochondrial DNA (petite strain) (1) and the other with a large peak of mitochondrial DNA strain (2). Strain (1) has been labelled with ^{3}H adenine and the tritium counts are shown as solid triangles while strain (2) was labelled with ^{14}C adenine and the carbon counts are shown by solid circles. (After Weisogel, 1971. *J. Biol. Chem.*, **246**, 5117.)

Table 4.2 Dissimilarity of mitochondrial DNA of two Amphibian species. Hybridization between *Xenopus laevis* and *X. mulleri* mitochondrial RNA and DNA*

Type of mitochondrial RNA	mtDNA	RNA hybridized ^{3}H:^{32}P ratio
rRNA	*X. laevis*	0.5
	X. mulleri	2.4
4SRNA	*X. laevis*	0.4
	X. mulleri	2.8
Complete RNA	*X. laevis*	0.22
	X. mulleri	22

* Each hybridization mixture contained ^{32}P-labeled *X. laevis* RNA and ^{3}H-labeled *X. mulleri* RNA. Filters containing mtDNA of *X. laevis* and filters containing mtDNA of *X. mulleri* were hybridized together in these RNA solutions. Reproduced with permission from Dawid (1972).

Dawid at Carnegie Institute in Washington (1972). Although the mitochondrial DNAs of the two species have the same molecular weight and base composition they have very different base sequences as shown by attempted hybridization of the two in a variety of conditions. Since these DNAs can be so readily distinguished by hybridization, Dawid has been able to study the mode of inheritance of the mitochondria in the interspecific cross between these species. The answer is unequivocal—whichever way the cross is made, the mitochondria are inherited cytoplasmically and maternally. A search for the presence of a nuclear copy of the mitochondrial DNA, has been unsuccessful. It seems safe to say that in *Xenopus*, and perhaps in most or all eukaryotes, the mitochondrial DNA is distinctive and is not represented in the nuclear DNA. In addition it now seems likely that in most or all vertebrates the mitochondria are exclusively derived from the mother (Hutchinson, *et al.*, 1974).

From the facts assembled in Table 4.1 it is hard to escape the conclusion that mitochondria have a direct evolutionary relationship to intracellular symbiotic bacteria. The fourth experimental system which yields information about mitochondrial role in differentiation does so by inference, since the units are not mitochondria but bacteria. These units are the alpha, kappa, and lambda particles found in some stocks of *Paramecium* and certain other ciliates (Fig. 4.14). Recently these (Gibson, 1970) have been identified as bacteria. This identification depends on observations of four kinds: (1) the presence of flagella in the

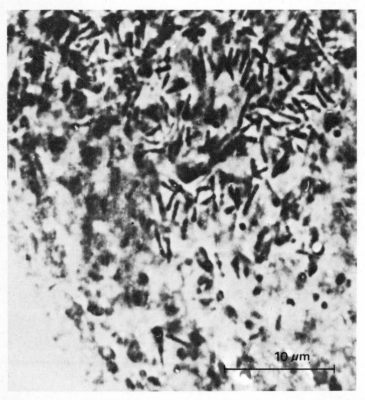

Fig. 4.14 Lambda particles in *Paramecium*. Part of *Paramecium* ,cell fixed in osmic acid vapour and stained with lacto-orcein. Lambda particles are clearly visible. (From Gibson, I. 1970. *Symp. Soc. exp. Biol.*, **24**, 379–401. Cambridge University Press.)

lambda particles (2) the presence of muramic acid (a bacterial cell wall constituent) in some of the particles (3) the ability to culture and maintain particles free of the ciliate cell (Van Wagtendonk *et al.*, 1963), and (4) the presence of cytochromes and glycolytic cycle enzymes within the particles.

We can afford space only to generalize about these interesting particles and the effects which they have on their hosts. Below is a summary of their more important characteristics:

1 Only some strains of any particular ciliate species normally possess the particles.
2 Single cells may harbour more than one type of particle.
3 The particles may be lost by rapid multiplication of the cell, the independent rate of particle replication lagging behind that of the cells.
4 Cells permanently harbouring the particles may exert some nuclear gene control over their presence, and loss or mutation of the nuclear gene or genes concerned leads to loss of the particles. Distinct nuclear genes govern maintenance of the separate types of particles.
5 Strains of *Paramecia* possessing kappa are often referred to as killer strains since they release a substance 'paramecin' which kills sensitive cells not possessing kappa. Sensitives may lose their sensitivity and become killers by acquiring kappa during conjugation. However, other strains are specifically killed by receiving kappa at conjugation and this is termed 'mate-killing' (Fig. 4.15).

For our purposes, these intracellular bacteria can be viewed as an early stage in the take-over process by the cell of the control of the symbiont. Since most of these particles can be cultured independently, it follows that their own genome remains relatively intact. In the case of mitochondria, the small size of the mitochondrial DNA unit and the presence within the nucleus of genes for most of the mitochondrial proteins, points towards a much more advanced stage of cellular control over the symbiotic unit. In differentiation, we can assume not only that modified expression of nuclear genes must often affect mitochondrial form and function but, more significantly, that such autonomy as mitochondria possess makes them an important factor in *determining* the differentiation of some cells.

(b) Chloroplasts

Just as mitochondria are now widely believed to be derived from intracellular symbiotic bacteria, so a strong case can be made out for chloroplasts originating from blue-green algae (see Figs. 4.16, Table 4.3). Like mitochondria, chloroplasts have their own unique DNA, ribosomal RNA, and some transfer RNA (Bell, 1970). Chloroplasts also demonstrate cytoplasmic continuity and there is no evidence that chloroplasts, once lost, can be resynthesized from other cell components.

However, although in the filamentous algae *Spirogyra* and *Nitella* there is good evidence for existing chloroplasts dividing and giving rise to daughter chloroplasts, this does not hold for higher plants. Here, chloroplasts are derived from a colourless primordium termed a proplastid, mature chloroplasts having lost the ability to divide and replicate.

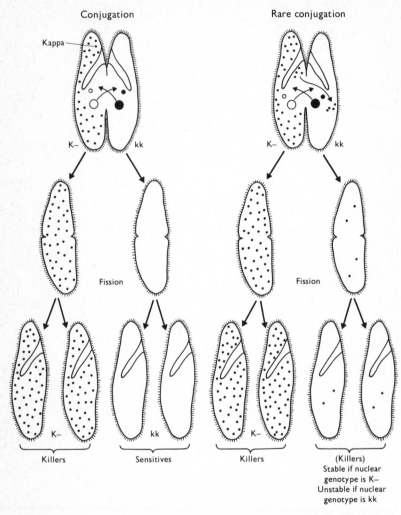

Fig. 4.15   Conjugation in Paramecium and the killer trait. Progeny of sensitives are killers only in rare situations where conjugation persists for a longer period so that kappa-containing cytoplasm is introduced into the conjugating sensitive. Kappa particles, however, are maintained only in the presence of a K – nuclear genotype. (After Burns, G. W. 1972. *Introduction to Heredity.* p. 371. Collier-Macmillan, London.)

Studies on the genetics of chloroplasts have been greatly facilitated by the use of plants which lack chlorophyll in the chloroplasts of some (variegated) or all (albino) of their cells. In many plants the chloroplast character is passed on both sexually and vegetatively, so that plant chimeras having both totally white and totally green shoots produce exclusively white or green cuttings and seedlings. Plants of diverse genera, for example *Epilobium, Nicotiana,* and *Zea,* all demonstrate such characteristics. This clearly implies that, in seed production, the egg provides all the chloroplasts for the future plant and the pollen parent does not sexually contribute chloroplasts to future generations. However, such a rule is by no means universal amongst plants, since in other genera, for example *Nepeta, Hypericum* and *Oenothera,* the pollen does contribute chloroplasts to the embryo and green seedlings may grow from the seeds of exclusively white shoots.

There are a number of aspects of chloroplast development and function very relevant to differentiation. One important observation, made on chloroplast inheritance in *Pelargonium,* is that the green plastids can replicate faster than the white, so that clones made between green and white plants yield a higher ratio of green plants whichever way the cross is effected. Preferential death of white embryos has been ruled out as a possible explanation. If the green and white plastids form sub-populations with differing replication times, it is not difficult to visualize a range of other chloroplast variants competing within cells and tissues in the same way, with differential survival of variants in different environments. Thus chloroplast variants other than simple absence of chlorophyll have been detected, such as altered grana structure in the evening primrose, *Oenothera* (Fig. 4.17). The altered chloroplasts appear to retain the feature as a constant genetic character.

As with mitochondria, some chloroplast proteins are coded for by the nuclear genome of the cell. Two chloroplast enzymes, malic dehydrogenase and lactate dehydrogenase, have been found to be coded in the nucleus of *Acetabularia* (Schweiger, 1970). In barley, a series of mutants are known in which chloroplast differentiation is incomplete due to defective proteins encoded in the nuclear DNA. So although plastid growth and replication may be independent of the nucleus, plastid differentiation and maturation are not. This leads us to consider perhaps the most revealing series of experiments on plastid control and development, again in the evening primrose, *Oenothera,* (Schötz, 1970). In hybrids between *O. lamarchiana* and *O. hookeri,* plants with pale plastids arise, but the paleness of the chloroplasts is temporary and dependent on the cellular environment. The tops of the cotyledons are green and the whole plant becomes green some time before flowering. It is clear, then, that the absence of chlorophyll in the chloroplasts is dependent on the expression of genes only in certain situations. These genes are almost certainly nuclear. Analysis of the fine structure of the chloroplasts in these plant tissues reveals that it is not simply presence or absence of chlorophyll which is at stake, but that the arrangement of the chloroplast membranes is abnormal. The precise relationship between the activity of specific genes and alterations in chloroplast fine structure remains to be resolved.

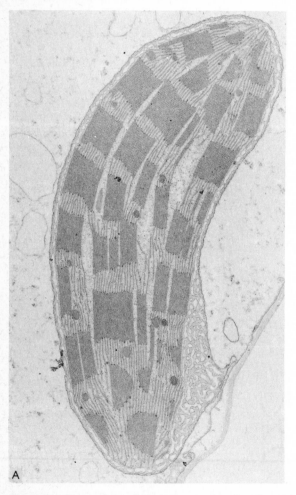

Fig. 4.16 Comparison of electron micrographs of a typical chloroplast (A), and the blue-green alga *Anabaena* (B). (A, from Burns, G. W. 1972. *Science of Genetics. An Introduction to Heredity.* Collier-Macmillan, London. B, from Leak, L. V. 1967. *J. Ultrastruc. Res.*, **20**, 190–205. Academic Press Inc. New York.)

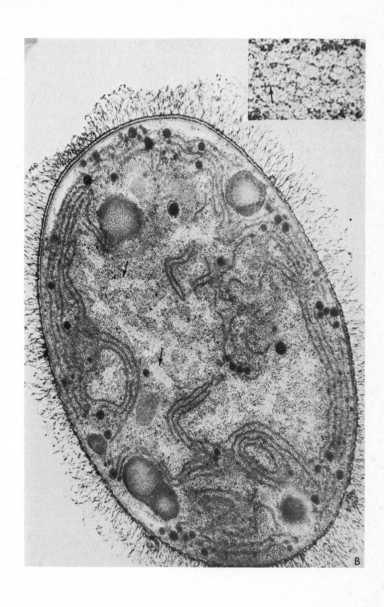

Table 4.3 Comparative biology of unicellular blue-green algae and higher plant chloroplasts (Echlin, 1966)

	Unicellular blue-green algae	Higher plant chloroplasts
Size	About 6 μm	4–8 μm
Cell wall	Present. Contains diaminopimelic acid	Absent
Osmotically active limiting membrane	Present. About 7 nm across	Present. About 7 nm across. Usually double
Ribosomes	Present	Present. Different from cytoplasmic ribosomes
Photosynthetic apparatus	Multilamellate. Peripherally arranged closed sacs. Each closed sac bounded by a 7–8 nm membrane	Multilamellate. Closed sacs are organized into stacks of grana lamellae interconnected by means of stroma lamellae. Each closed sac bounded by a 7- to 8-nm membrane
Photosynthetic mechanism	Photolysis of water with consequent evolution of oxygen. Primary mechanism coupled with photosynthetic phosphorylation	Photolysis of water with consequent evolution of oxygen. Primary mechanism coupled with photosynthetic phosphorylation
Primary photosynthetic pigment	Chlorophyll a plus accessory pigments	Chlorophyll a plus accessory pigments
Method of reproduction	Binary fission	Binary fission. Origin *de novo* from preexisting membranes within the cell. Origin *de novo* from preexisting proplastids within the cell
Genetic apparatus	Little known. Evidence suggests it is a non-Mendelian inheritance mechanism as seen in bacteria	Evidence suggests that it consists of a nonchromosomal heredity system with an absence of Mendelian segregation ratios and of linkage
DNA	Present as 2.5 nm fibrils; not organized into chromosomes. DNA not associated with histones	Present as 2.5 nm fibrils; not organized into chromosomes. DNA not associated with histones. In the few cases that have been investigated, chloroplast DNA has different base ratios from nuclear DNA
RNA	Present	Present. A messenger RNA complementary to chloroplast DNA has been found. A DNA-dependent RNA synthesis has been reported
Protein synthesis	Present	Present
Degree of autonomy	Entirely fre living	Intact chloroplasts prepared by improved isolation techniques appear to be able to perform at least all the photosynthetic functions normally ascribed to this organelle within the cell. No data available concerning other metabolic pathways

Table 4.3 *(contd.)*

	Unicellular blue-green algae	Higher plant chloroplasts
Product of photosynthesis	Present as cyanophycean starch which appears to be equivalent to the amylopectin of higher plant starches	Present as starch, but quickly translocated to other parts of the plant

Echlin (1966).

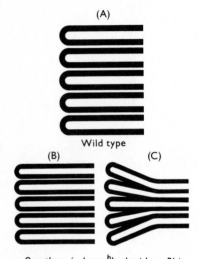

(A)

Wild type

(B)   (C)

Oenothera (velams. ^hhookeri lam.-Pl.)

Fig. 4.17 Organization of chloroplast grana. Diagram showing differences in the piling-up of the membranes to form grana.
(A) Wild type—harmonic co-operation between genome and plastome;
(B–C) types with disharmonic genome-plastome combinations;
(B, C) *Oenothera* (*lamarckiana* × *hookeri*) *velans·hookeri* with *lamarckiana* plastids.
(After Schötz, F. 1970. *Symp. Soc. exp. Biol.*, **24**, 48. Cambridge University Press.)

5

The role of hormones

It should now be clear that the factors which determine the extent and direction of differentiation are manifold and, especially during early development, are mainly active over very short distances or are directly transferred from one cell to another by intercellular contact. But as the multicellular organism develops in complexity and in size, it becomes necessary for homeostasis to be operative over much longer distances and to be responsive to changes occurring in distant organs.

We see such communication even within prokaryotic cells in the production by bacteria of colicins directed for offensive purposes against other bacteria, or in the remarkable anticipation of the important functions of cyclic AMP by its use as an aggregation factor in bringing together single cells of the slime mould *Dictyostelium discoideum* when they are subjected to starvation conditions. In eukaryotes, however, chemical communication is much enhanced and becomes the prerogative of hormones.

The range of substances which play a hormonal role in nature is very wide, and some are peripheral to our discussions. They extend from molecules which carry information between whole organisms, even of differing species, to those which are little more than on-site effector molecules such as acetyl choline at the neuromuscular junction. But we do well to recognize that there is no hard and fast line to be drawn anywhere between these two extremes, so that any definition of a hormone must be one of semantic convenience rather than a reflection of a natural grouping. In the broadest definition, a hormone is a substance which is produced by a living organism solely or partly to carry rather precise biological information. The classical hormone is manufactured at a precise site, often in a specific gland, and is active in discharging its information at a target organ elsewhere in the same organism. But with chalones, which affect the tissue of their origin, and pheromones which carry information from one organism to another, the classical concept must be extended.

It is proposed to discuss the role of hormones in differentiation under the following eight headings:

Pheromones
Hormone-like substances exchanged between species

Vertebrate steroid hormones
Vertebrate non-steroid hormones
Neurosecretory hormones
Insect hormones
Plant hormones
Chalones

This arrangement is purely of *ad hoc* value and must not be interpreted to have deeper significance. Vertebrate hormones are mostly divisible into two classes—steroids or polypeptides and amino acid derivatives. However, many overlaps in this classification come to mind. Thus the insect hormone ecdysone is a steroid and the plant hormone gibberellin is closely related to the steroids. Nevertheless for the purposes of discussing the implications of hormones for cellular differentiation, *this* subdivision will prove useful.

5.1 Pheromones

Pheromones are communication molecules operating between individuals within a species and most elicit a temporary behavioural response rather than a permanent differentiating change. Thus the discharge of traces of isoamyl acetate at the sting site by the honey bee excites other bees to discharge their stings in the same area. On the other hand the pheromone released by crowded individuals of the desert locust *Schistocerca gregaria* determines the degree of pigmentation and rate of development of adjacent individuals (Nolte, May and Thomas 1970), while oogenesis in worker honey bees (*Apis mellifera*) is inhibited by molecules secreted from the queen bee's mandibular glands (Butler and Simpson 1958). Similarly, reproductive development and its consequent caste determination is controlled in termites by pheromones produced by the already developed male and female reproductives. Pheromones are also released by some flowers as insect attractants and probably act as sex attractants in mammals. It is apparent, then, that pheromones, or exohormones as they were at one time termed, can play important roles in determining the degree and direction of tissue differentiation in certain species.

5.2 Hormone-like substances exchanged between species

There are some remarkable examples of the development of an individual of one species being affected by substances produced from another individual of a differing species. For example, ecdysone the insect hormone, or closely related compounds with similar activity, occur in high concentration in some plants, and are surely there because they afford protection to the plant by interfering drastically with the development of any insect unfortunate enough to feed on them. Plants containing such phytoecdysones include bracken fern and the European yew *Taxus baccata*. Certainly both plants appear to be strikingly free from insect attack. It is also likely that many animal parasites may alter hormone balance in their hosts in order to delay or prevent their metamorphosis. This is well established for the protozoan parasite *Nosema*, which blocks the juvenile hormone activity in its insect host and so prevents host metamorphoses (Fisher and Samborne 1964).

The researches of Miriam Rothschild (1965) have revealed an extraordinary relationship between the rabbit flea and the hormonal state of its rabbit host. Apparently the flea's sexual maturity is governed by the steroid hormones which circulate in the rabbit during pregnancy. Fleas, therefore, do not become sexually mature while feeding on nonpregnant rabbits. However, under the influence of the circulating steroids of the pregnant rabbit, the fleas reach sexual maturity, mate and lay eggs in the rabbit nest: the emergence of the young fleas coincides with the arrival of young rabbits, thus ensuring a fresh supply of host for the young fleas.

5.3 Vertebrate steroid hormones

Many of the important animal hormones are steroids, including oestrogen, androgen, aldosterone, progesterone and corticosteroid. They are small molecules with a tetracyclic nucleus and are related to other common biological molecules such as cholesterol and some bile salts. Being relatively small and lipid soluble they perhaps pass more easily through the plasma membrane of cells, and their action appears most frequently to involve the entry of the hormone into the target cell. Certainly the evidence for a direct implication of the hormone molecule at the level of DNA transcription would appear to be stronger for certain steroids than for any other hormones.

One system which has been investigated in some detail by O'Malley (1972) and his collaborators is the behaviour of chick oviduct when stimulated with oestrogen and progesterone. It seems probable that in this situation the hormones specifically alter gene transcription and induce altered RNA polymerase activity and the formation of novel nuclear RNA. But almost certainly the effects of any one steroid hormone on a target tissue are multiple and it appears that a single hormone may affect the different tissues in quite different ways. A bewildering array of diverse effects has been recorded for adrenal glucocorticoid hormone involving synthesis of specific proteins, inactivation of lymphocytes and inhibition of fibroblast growth.

Although some workers believe that cyclic AMP may be involved in the action of the sex steroids, the mechanics of steroid hormone influence on transcription probably normally involves the following pattern. Firstly, the small non-polar steroid molecule will pass fairly easily through the plasma membrane where it associates non-covalently with specific receptor molecules. Such specific binding proteins have been isolated from target tissues of many steroids. Once bound, the steroid-receptor complex moves to the nucleus where the complex may act as a specific activator of particular genes. But it must be emphasised that the precise way in which the steroid molecule interacts with the transcribing machinery of the cell is not known. One obvious problem is involved in the proposition that the steroid molecule acts either as a specific de-repressor or inducer. It will be recalled that in the induction of the galactosidase operon in *E. coli* very low concentrations of inducer molecule are sufficient to achieve full synthesis. Why then are steroid molecules needed in high concentration? When steroid molecules are radioactively labelled and applied to a target tissue, only a very small portion of the initial radioactivity is recoverable in the nuclear fraction. This itself is curious if the only primary site of action of the molecule is in the nuclear chromatin. But even the small amount of steroid which does apparently penetrate to the nucleus seems excessively large for inducing or derepressing a single gene.

We must conclude, then, that although steroids are better candidates than any other type of hormone for the role of primary effectors of genetic expression, their effects on target tissues are generally complex and their precise interaction with the DNA obscure.

5.4 Vertebrate non-steroid hormones

In the main, animal hormones which are not steroids are polypeptides, although some such as adrenaline, thyroxine and histamine are either amino acids or amino acid derivatives. The protein nature of insulin has been known for a long time and it was first crystallized by Abel in 1926; the determination of its amino acid sequence and of the mode of binding of its two polypeptide chains by Sanger in 1953 marked the first milestone in determining the structure of any protein. The insulin molecule has a molecular weight of 6000, while other polypeptide hormones such as the growth hormone STH (somatotropic hormone) have molecular weights of up to 48 000.

With the exception of molecules like adrenaline and thyroxine, the non-steroid hormones are very much larger than the steroids. We should not be surprised then, to find that the plasma membrane of the target cell constitutes a barrier to the polypeptide hormones, and their means of influencing the metabolism and destiny of the target cell is very different from that of the steroids. We have already determined that the steroid molecule enters the target cell and quite possibly acts on its genetic material in a direct and highly specific way. Not so the polypeptide hormones. Their mission ends at the cell surface, and the interpretation of their presence there and the conveyance of that information to the genetic material or other controlling centres of the cell must be carried out by a mechanism within the cell. In short, a transition exists between the extracellular signal and the intracellular action, and this is normally provided by what has come to be called the second messenger system (Fig. 5.1).

Due largely to the pioneering work of Sutherland and his collaborators in Tennessee, we now know that the second messenger in almost all cells is the cyclic AMP molecule, Adenosine 3′,5′-mono-phosphate, (Robinson, Butcher and Sutherland, 1971). On the inner surface of the cell membrane a hormone sensitive enzyme system, adenyl cyclase, is located which catalyses the formation of cyclic AMP from ATP (adenosine triphosphate), one of the main energy storing molecules of the cell. The importance of cyclic AMP in biology can scarcely be exaggerated, providing as it does the functional role which had puzzled biochemists and biologists for a decade—an information system operating between the cell surface and the cell's metabolic machinery. And such a link is central to the problem of differentiation. Although, as we have discussed in chapter 1, some differentiation may be an inbuilt and predetermined process, many of the permanent changes in developing cells and tissues are clearly in response to extracellular factors, whether it be contact with neighbouring cells or changes in the outer environment.

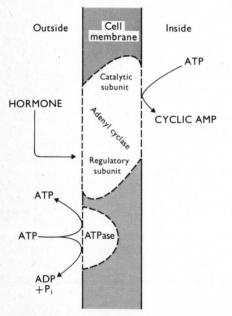

Fig. 5.1 The relationship between hormones, extracellular and intracellular ATP, and cyclic AMP in intact cells. (After Boutchen, R. W., Robison, G. A. and Sutherland, E. W. 1972 in *Biochemical Actions of Hormones*, Vol II. (ed.) G. Litwack. Figure 4, page 30. Academic Press Inc. New York.)

It is interesting to look at the incidence of cyclic AMP in nature. The molecule has been identified in almost all the tissues of multicellular animals so far studied as well as in many unicellular organisms. However, to date, neither cyclic AMP nor adenyl cyclase have been found in higher plants (Robinson, Butcher and Sutherland, 1971). When bacteria are grown on glucose-deprived media, it is cyclic AMP which promotes messenger RNA synthesis from genes whose products can relieve the carbohydrate deficiency. As mentioned on page 7, cyclic AMP has been identified as the aggregation factor in the slime mould *Dictyostelium discoidium* (Bonner *et al.*, 1969). In conditions of abundant food, this remarkable organism exists as single free amoebic cells. However, when food runs short, the single cells aggregate into a 'slug' which is capable of integrated movement. If food shortage continues, the slug becomes involved in a complex differentiation leading to the production of a sporing outgrowth. It is now known that the factor released by the starved amoebae, and which 'attracts' neighbouring cells to move towards and associate with it, is cyclic AMP.

Given, then that cyclic AMP is synthesized by adenyl cyclase at the cell surface in response to the presence of polypeptide hormone there, how does the cyclic AMP induce the array of cellular events which we recognize as the response to the hormone?

The first point to be made is that the cellular response to surface hormone is not necessarily an increase in levels of cyclic AMP. Some hormones, such as insulin, lead to decreased intracellular cyclic AMP and others, such as the catecholamines, may operate in different ways via either increase or decrease in levels of intracellular cyclic AMP (Table 5.1). We now know that in addition to the adenyl cyclase system, which controls the synthesis of the second messenger molecule, another enzyme system, the cyclic nucleotide phosphodiesterases, inactivate it by converting it to ordinary $5'$-AMP (Sutherland and Rall, 1958). How cyclic AMP is released from cells, as in the slime mould aggregation phenomenon, is not known. Although the evidence for the location of the enzyme making cyclic AMP in the cell membrane is now very strong, the normal siting of the cyclic AMP breakdown enzymes remains obscure. Clearly, if they are not resident in the plasma membrane, some communication system must be invoked to relay information from, say, insulin at the cell surface, and the cyclic AMP phosphodiesterases whose activity it stimulates.

Our understanding of the intracellular action of cyclic AMP stems from the elegant studies of Krebs and his co-workers in recent years (Krebs, Huston and Hunkeler, 1968). It seems likely that the cyclic molecule interacts in an allosteric fashion with phosphoprotein kinases. So we can see that, via cyclic AMP and the protein kinases, surface hormones will determine the levels of many intracellular enzymes and hence determine the metabolic state of the target cell. From this model we could predict firstly that different hormones might evoke the *same* response from a particular target cell, and secondly that the same hormone might evoke *different* responses in differing target cells. In fact both situations obtain and indeed the day to day response of the cells of a multicellular organism to its circulating polypeptide hormones is a highly complicated and ever changing process. But only by such a process can an organism hope to maintain homeostasis in the face of continual changes in the internal and external environment.

Table 5.1 Some hormone actions mediated by changes in cyclic AMP. (From Butcher, R. W., Robison, G. A. and Sutherland, E. W. 1972. In *Biochemical Actions of Hormones*, (ed.) G. Litwack. Vol. II, Table 1, page 22. Academic Press Inc., New York.)

Hormone	Tissue	Effect
Increased cyclic AMP levels		
Adrenocorticotropic	Adrenal cortex	↑Steroidogenesis
hormone	Fat (rat)[c]	↑Lipolysis
Luteinizing hormone	Corpus luteum, ovary, testis	↑Steroidogenesis
	Fat	↑Lipolysis
Catecholamines	Fat	↑Lipolysis
	Liver	↑Glycogenolysis, ↑gluconeogenesis
	Skeletal muscle	↑Glycogenolysis
	Heart	↑Inotropic effect
	Salivary gland	↑Amylase secretion
	Uterus	Relaxation
Glucagon	Liver	↑Glycogenolysis, ↑gluconeogenesis, ↑induction of enzymes
	Fat	↑Lipolysis
	Pancreatic β-cells	↑Insulin release
	Heart	↑Inotropic effect
Thyroid stimulating hormone	Thyroid	↑Thyroid hormone release
	Fat	↑Lipolysis
Melanocyte stimulating hormone	Dorsal frog skin	↑Darkening
Parathyroid hormone	Kidney	↑Phosphaturea
	Bone	↑Ca^{2+} resorption
Vasopressin	Toad bladder, renal medulla	↑Permeability
Hypothalamic releasing factors	Adenohypophysis	↑Release of trophic hormones
Prostaglandins	Platelets	↓Aggregation
	Thyroid	↑Thyroid hormone release
	Adenohypophysis	↑Release of trophic hormones

Table 5.1 (*contd.*)

Hormone	Tissue	Effect
Decreased cyclic AMP levels		
Insulin	Fat	↓Lipolysis
	Liver	↓Glycogenolysis, gluconeogenesis
Prostaglandins	Fat	↓Lipolysis
	Toad bladder	↓Permeability
Catecholamines (α-adrenergic stimuli)	Frog skin	↓Darkening
	Pancreas	↓Insulin release
	Platelets	↓Aggregation
Melatonin	Frog skin	↓Darkening

↑ = increase; ↓ = decrease.

One new development which should be mentioned here is the discovery of cyclic GMP (guanine monophosphate) in animal tissues (Steiner, Parker and Kipnis, 1970). It seems possible that cyclic GMP will also be found to be implicated in hormone action, but probably with a much more restricted role than cyclic AMP.

It should be emphasized that not all polypeptide hormones act via cyclic AMP, and some which do operate via that pathway have other actions as well. For example, insulin enhances protein synthesis by an apparently direct action on translation, and growth hormone stimulates amino acid transport. Both of these effects are independent of cyclic AMP.

Two other non-steroid hormones deserve mention here. The first, erythropoietin, is a glycoprotein complex of about 60 000 M.wt., which is secreted by the vertebrate kidney. Its mechanism of action is still poorly understood. It certainly stimulates the differentiation of erythropoietic stem cells into the erythroid series and perhaps it also regulates the rate of red blood cell production. One of its earliest effects on erythroid precursor cells is enhanced synthesis of very high molecular weight RNA (Gross and Goldwasser, 1969).

A second non-steroid hormone which merits comment here is prostaglandin. This substance, or rather group of substances, is present in most, if not all vertebrate tissues, but is found in high concentration in the seminal fluid of man and other mammals. Prostaglandins are cyclic, hydroxy fatty acids, and their primary physiological effect seems to be the stimulation of uterine smooth muscle, particularly at parturition. However, very many other diverse physiological effects have been observed in prostaglandin action, and some workers are unhappy about their being categorized as hormones at all.

5.5 Neurosecretory hormones

Neurosecretory hormones deserve special mention not because they are a special category of hormones or because their actions on target tissues are in any way different from hormones released from non-neural

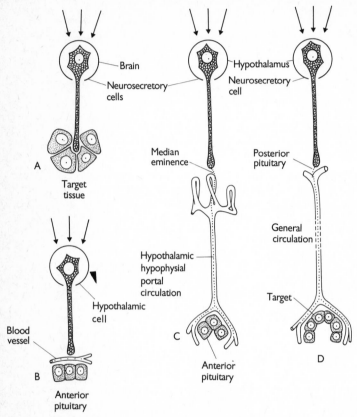

Fig. 5.2 Forms of neurosecretory cells and their connections. Arrows indicate neural and chemical messages from internal or external stimuli impinging on the neurosecretory cell (at the top) in each system. (a) Cells in the brain of an insect having their axon terminals in direct contact with target endocrine cells in the corpus allatum. This may be the most primitive arrangement; for in all other systems the secretion traverses some part of the circulation. (b) The shortest possible distance in the circulation separates the hypothalamic cell in the teleost fish from its target endocrine cells in the adenohypophysis (anterior pituitary). (c) Hypothalamic cells in the tetrapod brain secrete into a limited, but longer, portal circulation to reach a similar target. (d) Neurosecretory cells may also discharge into the general circulation by axon terminals forming neuro-haemal organs, as in the tetrapod neural lobe. (After Scharrer, E. 1965. *Archs. Anat. microsc. Morph. exp.*, **54**, 359–379.)

tissues. In fact hormones such as nor-adrenalin, produced by neuro-secretion, are also synthesized in the 'chromaffin' cells of the adrenal medulla and other non-neural tissue; though, of course, embryologically neural and glandular tissues have a common ectodermal origin. The significance of neurosecretion in relation to differentiation is that it indicates a partial shortcutting of an organism's response to its environment. A typical neurosecretory cell resembles a normal neuron, except that its axon terminates, not in a motor end plate or a synapse, but in an axonal bulb, often situated close to a blood vessel (see Figs. 5.2 and 5.3). The proximity of the axonal bulb of the neurosecretory cell to a blood vessel enables the cell to discharge hormone from the bulb, the hormone then passing through the wall of the vessel and so entering the circulation. Not all neurosecretory hormones rely on the blood circulation for transport to their target tissue. Some may pass directly from the terminal bulb of the neurosecretory cell to neighbouring target cells. So we can see

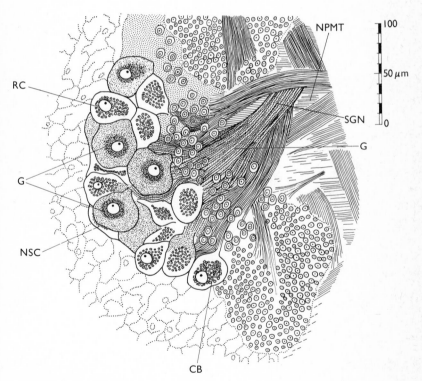

Fig. 5.3 Neurosecretory cells (NSC) in the ganglionic-X-organ of a crab, *Sesarma*. Stained granules of secretion (G) pass down the axons (SGN) to the sinus gland. Some cells (CB) have few granules, others (RC) are almost depleted of granules. Small ganglion cells and nerve fibres (NPMT) of the surrounding terminal medulla are also shown.

here a direct relationship between endocrine control of cellular metabolism and neuromuscular control, since acetylcholine transmits the
signal from the neurone at the motor end plate and induces the resultant
reaction of the muscle cell.

Most neurosecretory cells are located within neural tissue, and their
immediate juxtaposition with other nerve cells presumably speeds up the
inevitable delay between sensory input to the organism from the external
environment and release of hormone from the secretory cell. Apparently
the hormone is synthesized in the main cell body and carried down the
axon in combination with other proteins, often designated as carrier
proteins. This movement of hormone down the axon constitutes part of
the 'axonic flow' characteristic of neurons. Electron microscopy has
revealed that the hormone molecules and carrier protein are located
within neurosecretory granules of some 300 nm diameter and bounded
by membranes, and it is apparently these granules which cytologists
have observed for many years in the axons of neurosecretory cells
(Scharrer and Brown, 1962).

Neurosecretory cells have now been recognized in all classes of
vertebrates and in very many of the invertebrate phyla. The hormone
products seem to be almost invariably amines, peptides or polypeptides.

The significance of neurosecretion for differentiation is that strictly
neural tissue does wield considerable influence over the regulation of
development in many organisms. Moreover, in some examples of
neurosecretion, distribution of the hormone by the bloodstream is
by-passed and the neurosecretory cell releases the hormone in close
proximity to the target cell. An example of such a system is to be found in
many insects, where neurosecretory axons from the protocerebrum in
the brain release their secretion directly adjacent to the target cells of
the corpora allata. In this way the insect brain controls release of the
juvenile hormone from the corpus allatum.

5.6 Insect hormones

The endocrinology of insects is conveniently discussed separately, not
because insect hormones are not analogous to other animal hormones
but because insects have provided unique opportunities to the experi-
menter. These opportunities accrue from the considerable tolerance of
insects towards radical operations such as ligation, decapitation and
grafting of two individuals together, and also from the characteristics of
their developmental system. And so more is known about hormonal
regulation in insects than perhaps any other animal group except
mammals.

(a) Ecdysone and juvenile hormone

During development insects undergo a metamorphosis which involves
a series of sharply defined stages, one stage leading to another by the

moulting of the epidermis and cuticle. The splitting and rejection of the old cuticle is termed ecdysis. Some of these developmental stages are roughly similar to one another, so that the larva which emerges at ecdysis is not unlike its progenitor. However, other moults produce a greatly altered form, the larva now emerging as a pupa or the pupa as an adult. This process is controlled by two hormones. Moulting is initiated by a hormone called ecdysone, which is produced by the prothoracic glands. Ecdysone does not determine the emergent form after the moult but only the actual occurrence of moult. A second hormone, normally called juvenile hormone, is responsible for determining the type of moult involved and it is secreted by the corpora allata, paired glands normally situated behind the insect brain. When juvenile hormone is freely secreted, ecdysone initiates a moult which results in a relatively unaltered animal form. In the absence of juvenile hormone ecdysone will precipitate a moult which leads to a matured and radically altered insect form.

Release of juvenile hormone from the corpora allata is apparently controlled directly by the brain via nerve connections, but release of ecdysone by the prothoracic glands is directed by a group of neurosecretory cells in the brain which release ecdysiotropin or brain hormone. Both ecdysone and juvenile hormones have now been extracted in pure form and highly active analogues of them have been prepared synthetically. Ecdysone has been identified as a steroid (Huber and Hoppe, 1965) and juvenile hormone as a terpenoid (Bowers *et al.*, 1965).

Some astonishingly simple experiments have proved important in elucidating the roles of these two hormones. An experiment originally reported by Fraenkel in 1935, and now repeated yearly in many biological teaching laboratories, involves ligaturing the larvae of the blowfly (*Calliphora*) just behind the thoracic region with a cotton thread. Following this procedure, only the anterior part of the larva forms the characteristic hard sclerotinized puparium. The posterior portion of the larva behind the ligature remains soft and 'larval' due to lack of penetration of ecdysone from the prothoracid glands. An interesting development of this experiment provides a useful assay for ecdysone activity, namely the induction of puparium formation posterior to the ligature by the injection of ecdysone.

The blood sucking bug *Rhodnius* has provided an equally useful biological assay for juvenile hormone. As shown in Fig. 5.4 administration of the hormone to a late stage *Rhodnius* larva leads to the retention of larval characteristics in the adult form.

(b) Polytene chromosomes and puffing

A more direct approach to the role of ecdysone in differentiation is provided by a study of the effects of this hormone on the polytene chromosomes of dipteran flies. In these chromosomes, as we have seen earlier (chapter 2) light microscopy can be used to provide evidence of

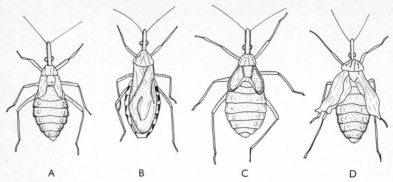

Fig. 5.4   Effects of juvenile hormone on the bug *Rhodnius*. (A) *Rhodnius prolixus* a normal fifth instar larva; (B) normal adult; (C) giant 'sixth instar' larva obtained by implanting a fourth instar larva's corpus allatum into a fifth instar larva; (D) a similarly obtained 'sixth instar' with larval cuticle on the abdomen, wings on the thorax, and other characters intermediate between larva and adult. (After Wigglesworth, V. B. 1954. *The Physiology of Insect Metamorphis*. Cambridge University Press.)

gene activity. These giant structures consist of chromosomes replicated 'sideways' many times, all the replicated strands lying in parallel and with more than one thousand times the haploid amount of DNA. Figure 5.5a shows a squashed preparation of these giant chromosomes which are characterized by a specific linear sequence of chromatic bands. The salivary glands of the fruit fly *Drosophila* are among the best known sources of these preparations. In light microscopy, polytene chromosomes can be seen to display regions of local distension termed puffs or in the case of the giant nucleolar puffs of *Chironomus*, Balbiani rings. As discussed in chapter 2 there is good evidence that a puff represents an active gene and administration or absence of ecdysone in the glandular tissue causes specific changes in the puffing pattern. Since the gene activation displayed in puffing involves localized RNA synthesis, this ecdysone-activated gene expression can be measured by local incorporation of labelled uridine. However, as with the steroid hormones of vertebrates, evidence for a direct effect of ecdysone on the genetic material remains speculative and its mode of action at a molecular level unclear.

(c) Imaginal discs

The differentiation of imaginal discs has been discussed briefly in chapter 2. However, the relationship of ecdysone to the growth and development of these discs is of interest to us here. The reader will recall that imaginal discs are groups of cells, of epidermal origin, which persist in a relatively undifferentiated but apparently committed state during larval growth. At metamorphosis, they are transformed into specialized adult structures. The weight of evidence points to a strong ecdysone dependence for imaginal disc differentiation and, as with puffing in

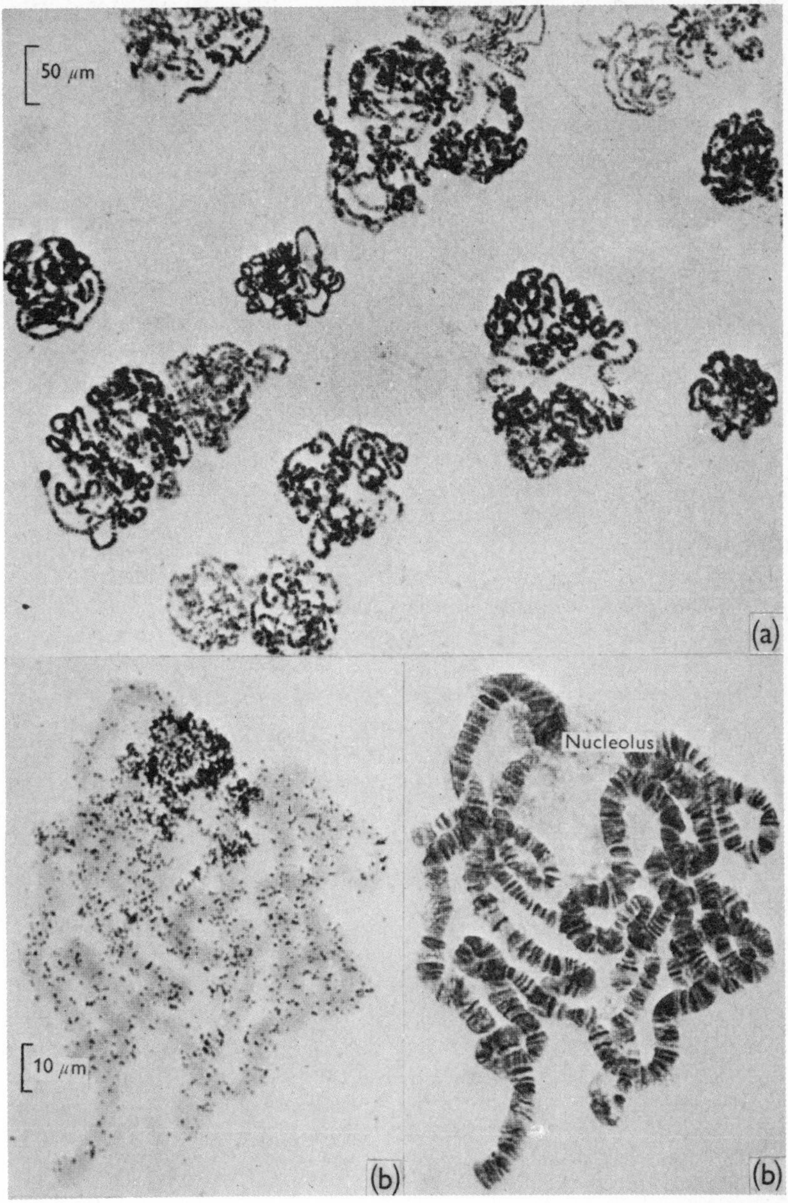

Fig. 5.5 *Drosophila* polytene chromosomes. (a) Squash preparation of a drop of a suspension of isolated nuclei of *Drosophila hydei*. (b) Distribution of tritiated uridine triphosphate incorporation after 20 min of *in vitro* incubation of the nuclei. (From Berendes, H. D. 1971. *Symp. Soc. exp. Biol.*, **25**, 145–61. Cambridge University Press.)

polytene chromosomes, this points to a key role for ecdysone in control-ling morphogenesis and cell differentiation in insects.

5.7 Plant hormones

The so-called plant hormones are generally less specific in their action than are animal hormones and in some ways are analogous to the chalones which control the growth of certain animal tissues. They are of various kinds, which we will discuss under the headings of auxins, gibberellins, cytokinins, and ethylene. However, since they frequently interact in producing their effects, it is not easy to be certain about discriminating between them when analysing their roles in plant cell differentiation.

(a) Auxins

Auxins were the first plant growth substances to be studied. Following their discovery in 1927, and after many problems associated with separat-ing and isolating active fractions from plant material, auxin has now been identified as indole-3-acetic acid (IAA). The comparatively simple structure of this molecule has enabled a range of synthetic auxins to be manufactured, some being utilized as important herbicides. Indeed the synthetic auxins MCPA (1-methyl-2 chlorophenoxyacetic acid) and 2-4-D (2-4-dichlorophenoxyacetic acid) are far and away the most widely used herbicides in the world today and are apparently relatively non-toxic to other organisms.

The effect of auxin on plant tissues are somewhat variable, and include root formation, stem elongation, inhibition of lateral bud development and inhibition of the abscission of leaves and fruits (Miller, 1970). A bio-assay for the activity of auxin has been developed in the form of a curvature test on the coleoptile, the sheath surrounding the first leaves, of the common oat *Avena*. In this test (Fig. 5.6), the shoots are decapitated and the shoot tips replaced by agar blocks containing the test solution. The blocks are placed on top of the cut surface, but on one side only, and the presence of auxin is betrayed by enhanced elongation of the side of the tip below the agar block. Such asymmetrical elongation of the shoot leads, of course, to a progressive curvature of the stem. It is interesting to recall that Darwin carried out many experiments on the responses of shoot tips to light and his finding was that covering the extreme shoot tip prevents light-initiated curvature in the entire stem. These experiments are discussed in his last book *The Power of Movement in Plants*, published in 1880.

Auxin induces cell enlargement in many plant tissues and cell division in some. The extremely rapid response of the *Avena* coleoptile to auxin is entirely attributable to cell enlargement and reaches optimal proportions within 90 minutes. There is no evidence that IAA actually enters the cell: rather it acts in a way analogous to the animal polypeptide hormones,

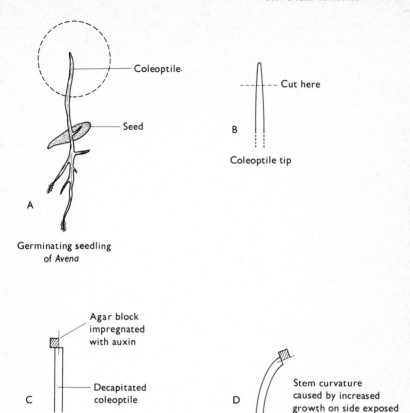

Fig. 5.6 *Avena* coleoptile test for auxin.

relying on a second messenger for its action. The second messenger involved in IAA activity is not known.

(b) Gibberellins

This class of plant hormone was first discovered in Japan between 1930 and 1940, but was not known in the western world until an English translation of the Japanese work became available after the second world war. Gibberellins are structurally similar to steroids and all possess the gibbane molecular skeleton. They are commonly identified by a bio-assay which involves the ability of the hormone to induce elongation of the stem in normally dwarf plants (Table 5.2). There are at least twenty different natural gibberellins with slightly differing physiological activities. No plants have been found which possess all the known gibberellins, but most plants have several. As with some other

Table 5.2 Comparison of some of the biological effects of gibberellins and auxins. (From *Physiology of Plant Growth and Development* by M. B. Wilkins (ed.) — 1969. McGraw-Hill Book Company (UK) Limited. Used with permission.)

Response	Effect of:	
	Gibberellin	Auxin
Dwarf pea stem growth, sections	n.e.	promotes
Dwarf pea stem growth, intact	promotes	n.e.
Cucumber hypocotyl growth, intact	promotes	promotes
Root growth	n.e.	inhibits
Parthenocarpic fruit growth, tomato	promotes	promotes
Cell division, tobacco pith	n.e.	promotes
Root initiation	n.e.	promotes
Flower initiation, long-day plants	promotes	n.e.
Seed germination, Grand Rapids lettuce	promotes	n.e.
Sex expression, cucumbers	promotes maleness	promotes femaleness

n.e.; no effect or only weak response.

plant growth regulators, the gibberellins seem to be active at their site of synthesis, although activity consequent upon their transport up or down the plant is known. Movement may be both by diffusion through the tissues and via the conducting vessels of the plant.

As with steroid hormones, the gibberellins probably act at the gene level after entering the target cell. It is interesting to note that a gibberellin preparation has been found to have some activity in a locust moulting assay, while the insect hormone ecdysone, also steroid-like, induces growth in an assay on dwarf peas. Surely an example of convergent evolution of growth regulating molecules?

(c) Cytokinins

Studies on the biochemical control of differentiation in plant tissue cultures by Professor Folke Skoog at the University of Wisconsin led to the discovery of a new group of plant growth regulators. The first to be isolated was termed kinetin, and this molecule, along with many very

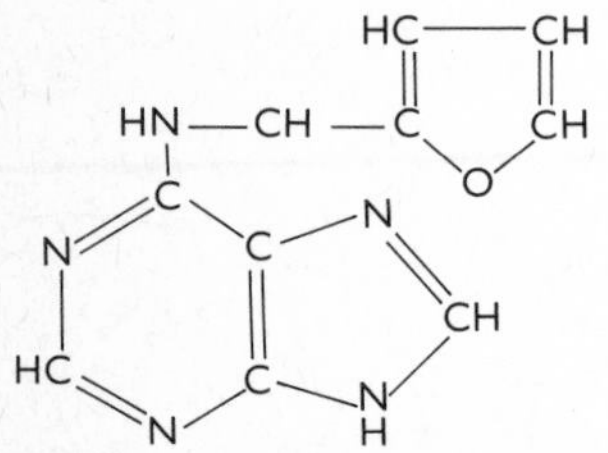

Fig. 5.7 Structure of kinetin.

similarly structured molecules, are known as cytokinins. Kinetin is a dehydrated deoxyadenosine (Fig. 5.7). Changes in the side chain of the molecule have little influence on its properties but alterations in the adenine moiety either eliminate or greatly reduce its biological activity. Cytokinins are defined as molecules which promote cytokinesis (cell division) in plant cells. They have also been found to promote cell division in protozoan cultures and cell cultures of tissues from higher animals.

Effects of cytokinins are diverse, but one particularly dramatic response is the differentiation of cultures of plant callus tissue to form buds or roots with appropriate concentrations of kinetin. The required concentrations of the active molecule are astonishingly low, one part per 10^9 producing a response. Research on cytokinins is at present in a very active state and some of the evidence relating to their mode of action frankly contradictory. It appears that the action of cytokinins involves their incorporation into some fractions of RNA, especially transfer RNA, and it is certainly appropriate to recall that there is good evidence that purines alkylated at the number 6 position (as are the cytokinins) lie next to the anticodon in several transfer RNAs from a variety of organisms. Perhaps then the differentiating effects of the cytokinins are caused by their influence on the active population of transfer RNA molecules within the affected cell.

(d) Ethylene

Ethylene $(CH_2\!\!=\!\!CH_2)$ is surely the most remarkable of the plant growth regulators, both because it is a gas at normal temperatures and pressures and because it is easily the simplest organic molecule with specific biological activity, having a molecular weight of only 28. It is highly active in very low concentrations and is 60–100 times more effective than the next most active molecule, propylene. Ethylene affects the ripening of fruit, leaf abscission, flowering and general growth regulation in plants, and the gas is believed to owe its activity to the induction of changes in cell permeability (Abeles, 1972). Green bananas placed in a dish beside riper yellow bananas will ripen rapidly due to the release of ethylene from the riper fruit. The relationship of the effects of ethylene to those of other plant regulators is still confused, since such molecules as auxins do induce the release of ethylene.

(e) General

In recent years it has become apparent that plant development and differentiation are controlled by the simultaneous involvement of several groups of molecules, probably often interacting. Moreover, the effects of the hormones which we have discussed are often further modified by naturally occurring inhibitors, including flavonols, phenols and abscisic acid. The evidence on the entire field is still confusing and muddled, often because identification of active fractions depends on bio-assays which fail to distinguish clearly between different types of growth regulators. Be that as it may, it is clear that plant differentiation is largely hormone controlled. Moreover, since by the nature of plant material, most cellular changes are permanent, we do not have to disregard in our study of plant cell differentiation the many temporary and reversible changes in tissue state or organ size characteristic of animals. Even the effects of auxin on cell size in the stem curvature assay may determine the permanent shape of the mature plant. Most plant hormones appear to be active at or close to the site of synthesis; some, however are transported to the target tissue, and presumably ethylene, being gaseous, may well be released by one plant and affect the growth and development of neighbouring plants.

The remarkable differentiation displayed by the phase changes of the ivy plant *Hedera helix* is also relevant here (see Table 5.3). Although there is no evidence that these phase changes are entirely controlled by plant hormones, reversion of the adult phase to the juvenile phase in rooted adult cuttings has been induced by treatment with gibberellic acid (Robbins, 1957). So here again, even if only in the capacity of a trigger factor, a plant hormone can be responsible for profound alterations in the phenotype of a plant.

Table 5.3 Distinguishing characters of juvenile and adult ivy (*Hedera helix*). (From Wareing, P. F. 1971. *Symp. Soc. exp. Biol.* **25**, 323–44.)

Juvenile characters	Adult characters
Three or five-lobed, palmate leaves	Entire, ovate leaves
Alternate phyllotaxy	Spiral phyllotaxy
Anthocyanin pigmentation of young leaves and stem	No anthocyanin pigmentation
Stems pubescent	Stems glabrous
Climbing and plagiotropic growth habit	Orthotropic growth habit
Shoots show unlimited growth and lack terminal buds	Shoots show limited growth terminated by buds with scales
Absence of flowering	Presence of flowers

5.8 Chalones

Chalones can be defined as tissue specific mitotic inhibitors which are active in determining and controlling organ size. Recognition of the humoral circulation of such growth regulators in animals is not new, and derives largely from a classic experiment carried out in 1951 by Bucher, Scott and Aub. In this experiment, two rats were anaesthetized and made parabiotic by joining their circulatory systems. Part of the liver of

Fig. 5.8 Effect of gibberellin on the growth of dwarf bean plants. Application of 20 μg of gibberellin has been made to plant on the right.

one rat was then excised. It had already been established that such partial hepatectomy leads to rapid liver regeneration with increased mitotic rate in the operated animal. What Bucher's experiment showed was that such regeneration and increased mitotic rate occurred also in the normal animal linked parabiotically to a partially hepatectomixed one. The evidence for a humoral agent specifically affecting the mitotic index of the liver seemed very strong. Although the experiment has been successfully repeated many times, the humoral factor has proved extremely resistant to satisfactory extraction and purification. Of course other interpretations of the phenomenon can be advanced which do not imply chalone activity.

Control of organ and tissue size by chalones has been consistently argued for by Professor William Bullough of London University and he and many other workers elsewhere have now produced evidence for such chalones being involved in the growth of epidermal and kidney tissue as well as in liver (Bullough, 1967; Thornley and Laurence, 1975). Such extraction and purification of chalones as has been achieved results in the hormonal activity being associated with a glycoprotein of relatively low molecular weight. The model of chalone activity normally proposed is that a cell or tissue will constantly release chalone molecules which, in high concentration, inhibit mitosis in that cell or tissue. Just how such inhibition is accomplished remains obscure, but ideas of a specific mitosis operon have been advanced, with the suggestion that the chalone would directly or indirectly repress the operon. Such suggestions are, to date, pure speculation.

When an organ such as the liver has reached its normal size, the concentration of liver chalone in circulation is deemed to be sufficiently high to prevent any further liver growth, and permit only enough mitosis to compensate for cell death. Circulating chalone is held to be unstable and so constant replacement is necessary to ensure continued inhibition of mitosis. If part of the liver is removed surgically, liver chalone production would be reduced, circulating liver chalone levels would fall due to normal chalone breakdown, and the remaining liver tissue, now exposed to reduced circulating chalone concentration, would rapidly burst into increased mitotic activity. When enough liver had regenerated to bring the humoral chalone back to a normal level, liver growth would cease. It is difficult to find any equally satisfactory alternative scheme for controlling organ size in animals, and such self regulation seems to make good sense. Presumably it is the inherent instability of chalones that renders their isolation so difficult.

We can conclude, then, that it is likely that many or most animal tissues regulate their own growth and development by production of self limiting hormones which, by their instability, permit rapid adjustment to partial loss of tissue. But much more information is needed about the interplay between chalones and other growth factors, especially during development.

Table 5.4 Model systems of hormone-dependent cell differentiation. (From Turkington, R. W. 1971, in *Development Aspects of the Cell Cycle*, (eds.) I. L. Cameron, G. M. Padilla and A. M. Zimmerman. Table II, p. 327. Academic Press Inc., New York.)

System	Differentiated cell	Marker protein(s)	Hormonal inducer(s)
Chick retina	Neural retina cell	Glutamine synthetase	Hydrocortisone
Chick oviduct	Tubular gland cell	Lysozyme, ovalbumin	Estrogenic hormones
	Goblet cell	Avidin	Progesterone
Bone marrow	Erythrocyte	Hemoglobin	Erythropoietin
Mammary gland	Alveolar cell	Casein, lactose synthetase	Insulin + hydrocortisone + prolactin
Insect larva	Cuticle	DOPA decarboxylase	Ecdysone

5.9 General comments

We have now reviewed the important role which hormones play in the growth, development and differentiation of plants and animals. Since many of the cellular responses to hormones are highly specific, biologists and biochemists have sought to use hormone-initiated protein synthesis as a model for the study of transcription, translation, and the control of genetic expression. Table 5.4 lists some of the most important examples of these systems, together with the particular proteins produced by the cell under hormonal influence which can be used as indicators of differentiation or altered genetic expression.

All these systems hold great promise, but also share common weaknesses. As we have discussed in this chapter, it is not as yet certain that any hormone operates by activating an individual cistron, directly or indirectly, and most hormones appear to have rather diverse influences on their target cell. Although the marker proteins listed are certainly synthesized under the influence of the hormones listed, in probably all cases their production can occur in some circumstances without hormonal interference.

6

Episomes, viruses and abnormal genetic elements

So far in this book our consideration of differentiation has been confined to the situation in the normal cell. Clearly cells differentiating in pathological situations will often display drastically altered patterns of growth and development. It is not always easy, however, to sharply categorize normality and pathology. Strains of bacteria harbouring prophage lambda, and clones of plants harbouring intracellular viruses, are instances of living systems in which the 'pathology' must be accepted as the 'normal' state. In addition to these examples, some genetic systems display features which do not fall into line with conventional patterns and are, in certain ways, reminiscent of bacterial episomes. In this chapter we will therefore discuss an assortment of examples which are either episomes, or viruses with a very stable relationship with the host cell, or genetic elements with an apparent kinship to episomes.

6.1 Episomes and plasmids in bacteria

Bacteria harbour a variety of relatively stable hereditary units which are or can be physically separate from the main genome or chromosome of the cell. These include the sex factor (F), colicin producing factors (Col), some coliphages such as P1, and R factors which carry certain drug resistance determinants. These extrachromosomal factors are loosely termed plasmids (Fig. 6.1). If they are known to be capable of existing in a state of integration with the host cell chromosome as well as autonomously they are then designated as episomes (Fig. 6.2). Since some parallels will be drawn between bacterial plasmids and certain genetic factors in higher cells, it is necessary to briefly review the biology of these hereditary units (see also Clowes, 1972).

All bacterial plasmids are composed of double stranded DNA which replicates out of phase with the host cell. Their relative abundance in the cell varies, the R factor in *Proteus*, for example, accounting for between 12 and 33% of the total cellular DNA (Novick, 1969). It is particularly relevant to the topic of differentiation to note that plasmids may greatly influence the characteristics of the host cell. Thus, some plasmids carry important bacterial genes and are suggestive of 'escaped' parts of the bacterial genome. Others, like the Col factors, induce the syntheses of

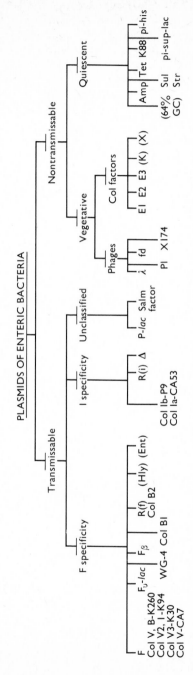

Fig. 6.1 Classification of plasmids harbored by enteric bacteria. Parentheses for Hly, Ent, and 64% G + C indicate that transmissibility is not certain; those for Col K and Col X indicate that there is reason to question their compatibility. The plasmids grouped as 'quiescent' are only inferred to be so — direct information is not available. (After Novick, R. P. 1969. *Bacteriological Revs.*, **33**, 2, 210–63. American Society for Microbiology.)

colicins, which are proteins capable of inducing lysis of other bacteria by attachment to the surface membranes. The sex factor, F, has a particularly crucial role in the conjugation of the strains of bacteria which harbour it, while the R factors confer resistance to penicillin and other drugs, phages, and ultra violet light.

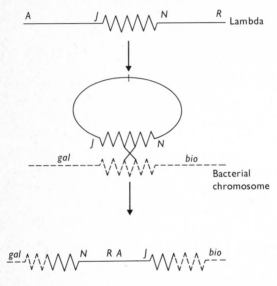

Fig. 6.2 Integration of phage lambda into the bacterial chromosome. (From Campbell, A. M. 1969. *Episomes*. Fig. 5.1. p. 69. Harper and Row, New York.)

In summary, bacterial plasmids are dispensable genetic factors which, when present, endow the host cell with new characteristics, some disadvantageous, some beneficial. Some plasmids are episomes, capable of integration into the host chromosome; in this state the episome will be replicated along with the chromosome and may not betray its presence by any phenotypic effect. The stable equilibrium of an integrated bacterial episome has a small chance of breaking down at any time, but can be easily terminated by U.V. radiation. In the unintegrated phase, the episome will often freely replicate in the cell and, as in the case of phage lambda, may induce lysis of the host cell.

The situation in which bacteria most nearly approach a process of differentiation is in sporulation (see chapter 1) and it is possible that episomes may be involved in the induction of that event (Rogolsky and Stepecky, 1964; Campbell, 1969). But whether bacterial sporulation ever involves episomes or not, it is important to recognize the possible parallelism shown by some aspects of higher cell differentiation.

6.2 Virus infection and latency

The possibility that many eukaryotic cells harbour latent virus integrated into the genome is an exciting and worrying possibility but one not easily investigated. Such a model for the persistence of oncogenic virus has often been suggested and will be discussed in the context of cancer in chapter 8. But certain examples of non-oncogenic viruses have some bearing on differentiation.

The first is herpes simplex virus. This virus, a large double stranded DNA virus, occurs in the nuclei of cells of infected persons (Fig. 6.3). Infection most frequently takes place in early childhood, and once infected, the virus normally persists throughout life. One of the most fascinating aspects of herpes simplex infection is that no pathological symptoms are apparent for most of the time: only after excessive exposure to sunlight or ultra violet light, or during an accompanying influenza or cold virus infection, does a temporary lesion appear on the facial skin.

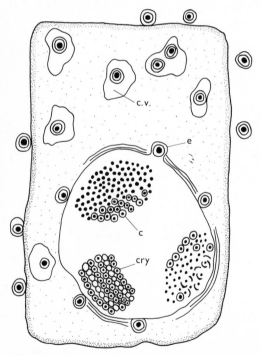

Fig. 6.3 Diagram of a cell showing herpesvirus replication. Viral DNA and the icosahedral capsids (c) form in the nucleus, and sometimes the capsids are arranged in a crystalline array (cry). Capsids acquire an envelope (e) by budding through the nuclear membrane and are then found in cytoplasmic vacuoles (c.v.) from which they are eventually released by egestion. (After Bernhard, W. 1964. in *Ciba Foundation Symposium on Cellular Injury*. Fig. 5, p. 220.

This lesion appears to betray vegetative growth of the herpes virus, but it rapidly subsides to leave, once more, normal and apparently healthy facial skin. Although it is known that the herpes simplex DNA is never truly integrated into the genome of the host cells, its latency and persistence during the quiescent phase is remarkable, particularly since, during latency, the virus persists, not in the skin cells, but in cells of the dorsal root ganglion serving the skin area. This virus makes us aware of the possibility that extra chromosomal infective units may reside indefinitely in tissue cells, only influencing cellular growth and metabolism for critical periods.

It is uncertain how many other viruses and viral-like factors may persist in animal tissues in a latent condition. One other which has been known for many years is the sigma virus occurring in some strains of *Drosophila melanogaster*. The presence of the virus was originally manifested in the laboratory by the extreme sensitivity of these strains to the paralyzing influence of CO_2, and such CO_2 sensitive strains were found to be common in nature (Kalmus *et al.*, 1954). The virus can be transmitted by injection of CO_2 resistant strains with extracts prepared from sensitive flies; once infected the virus is passed on to the progeny via the germ cells. Actually, states of differing stability of the virus exist, as revealed by the mode and efficiency of transmission from parents to offspring, and this has been interpreted to imply differing levels of viral integration in the insect tissues (Preer, 1971). Although attempts to isolate the virus have so far been unsuccessful, it has been visualized in the electron microscope within the tissues of sigma-bearing flies.

Turning now to plants, it is immediately obvious that plant tissues are much more susceptible to persistent viral infection than are animal tissues. This fact presumably reflects the absence of an immune defence mechanism in plants. It also implies that viruses may play a much greater role in affecting the development and differentiation of plant cells than animal cells. Thus entire clones of plants owe much of their characteristic morphology to persistent viral infection. For example, both the variegated *Abutilon* used extensively in ornamental plant bedding in parks, and the flowers displayed by some tulip fanciers, depend on viral-induced colour breaks. Such viral infections profoundly modify the expression of the genome in the production of the phenotype.

6.3 'Abnormal genetic elements'

Studies in plant genetics have also revealed some curious phenomena which, although providing no direct hint of viral involvement, are in some ways suggestive of bacterial episomes. Two outstanding examples are the high mutability genes discovered in maize and *Antirrhinum*.

Dr. Barbara McClintock has devoted some 25 years of careful study to the elucidation of highly mutable alleles in maize (McLintock, 1951). Her main findings are that certain specific gene loci demonstrate very

high mutability, many thousands of times higher than the normal mutation rates for other genes. Such loci appear to be regulated by 'controlling
elements' which can occupy different gene loci and are, indeed, frequently transported from one locus to another. Two mutually exclusive
sets of mutable alleles are present in maize, controlled by two quite
separate transportable elements, named by McClintock, Activator (Ac)
and Suppressor-mutator (Spm). Also of great interest is the observation
that certain phases of development are characterized by very high
mutational activity in these loci—for example late in the development
of the maize endosperm (Fig. 6.4).

Studies on the plant Antirrhinum by Fincham (1970) and his colleagues
have revealed very similar gene loci affecting pigment and pattern formation in the flower. Fincham quotes mutation rates in some Antirrhinum
tissues in which over 50% of the cells in the appropriate tissue have
undergone the particular mutation, and several per cent in the germ
cells. Once isolated, these mutations appear to be stable and to show a
regular mendelian segregational behaviour.

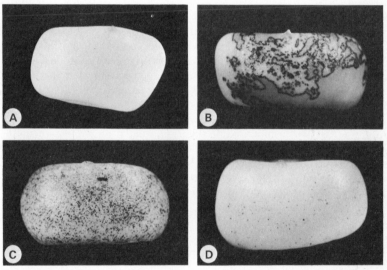

Fig. 6.4 Effect of *Activator* Gene on the action of *Dissociation* Gene *D* in maize.
(A) No *Ac* is present. The kernel is colorless due to the continued presence of
Ds, which inhibits the action of a nearby pigment-producing gene. (B) One *Ac*
gene is present. Breaks at *Ds* occur early in kernel development, leading to large
colored sectors. (C) Two *Ac* genes are present. Time of *Ds* action is delayed,
producing smaller sectors which appear as specks. (D) Three *Ac* genes are
present. *Ds* action is so delayed that relatively few and tiny specks are produced.
(From McClintock, B. 1951. in *Cold Spring Harbour Symposia on Quantitative
Biology*, **16**, 13–47. Figs. 10, 12, 14, 15.)

Caution must be exercised in interpreting these fascinating results and in drawing parallels with other biological and genetic phenomena. Nevertheless, in both maize and *Antirrhinum*, but particularly in the former, the genetic elements responsible for the gene repression mutation behave as if they were dispensable to the tissue and organism and are not, therefore, parts of the primary genome. Such a view of these factors, shared by both McClintock and Fincham, is certainly fairly close to the concept of a bacterial episome.

Two other biological systems provide relevant examples. These are the 'white' gene in *Drosophila* and the genetics of immunoglobulins. In the first example, the mutation rate of the 'w' gene was greatly affected by its position in the genome, apparently through association with a mobile genetic element adjacent to it. This genetic element was also found to be capable of excision from its locus adjacent to w and insertion elsewhere, sometimes carrying part of the 'w' locus with it (Green, 1969).

The genetics of immunoglobulins provides many thorny problems, not least the great diversity of the genes for the 'variable' portions of the immunoglobulin chains and their association, during transcription, with the genes for the 'constant' portion. Gally and Edelman (1972) have suggested that a special genetic organization, the 'translocon', might be involved, in which a chromosomal segment with both the relevant v and c genes juxtaposed would result from somatic translocation of the v gene from its original site and its insertion next to the appropriate c gene.

In concluding this chapter, let us attempt to draw the strands of evidence together. Bacteria harbour a range of genetic elements which profoundly affect their phenotypic characteristics but are actually foreign to the genome and may or may not be integrated into it. The tissues of eukaryotic organisms harbour some viruses which behave in a somewhat similar fashion and, in addition, some genetic phenomena in higher organisms show many of the characteristics indicative of foreign genetic elements. In observing and interpreting the expression of genes in differentiation, it is necessary to remember the fact that not all genes may be native to the organism. Applications of this concept to some types of cancer are discussed in chapter 8.

7

The cell surface and cell contact

The cell surface serves many important functions in relation to differentiation. It is the area of the cell in direct contact with the outer environment, including other cells, and is responsible for 'sieving' the molecules and monitoring the 'information' entering the cell from the outer environment. Cell surfaces also play a crucial role in cell movement and, since cellular environment and cellular position are key factors in determining the fate of a cell, it follows that some knowledge of the form and function of the cell surface is very necessary for a proper understanding of differentiation.

The living surface of almost all cells, whether bacterial, plant or animal, consists of a membrane, often termed the plasma membrane. Although it is certainly a highly dynamic structure, electron microscopy of fixed cells reveals a remarkably uniform unit membrane structure. This unit structure, appearing as two electron dense lines with a less dense space between, and with an overall dimension of 7·5 to 8·0 nm, consists essentially of two layers of protein covering a bimolecular middle layer of phospholipid (Fig. 7.1). Originally proposed by Davson and Danielli (1952) such a structure not only satisfies the electron microscope observations but also the known permeability and low surface tension properties of the cell surface. A modern view of the structure of the plasma membrane will be found in the short review by Bretscher and Raff (1975).

Although many animal cells present the plasma membrane as their outer cell surface, many other organisms possess other inert outer layers. Thus bacteria and plant cells are enclosed with inert layers of cellulose, mucin and other polysaccharides. Such cell walls are highly permeable lattice-like structures with skeletal functions. Indeed, the cells can often be grown in culture without these cell walls, but the normal shape of the cell is then lost.

Three aspects of cell surface phenomena are relevant to the problems of differentiation. They are cell to cell contact, cell locomotion, and cellular response to extracellular molecules at its surface.

7.1 Cell to cell contact

Even a fairly superficial look at electron micrographs of tissue cells reveal that not all adjacent areas of neighbouring cells are actually in contact,

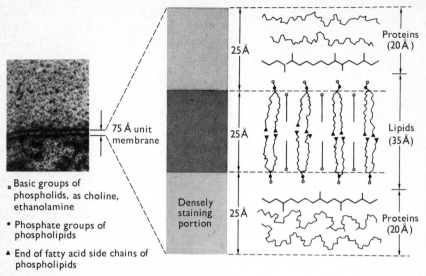

Basic groups of
phospholids, as choline,
ethanolamine

Phosphate groups of
phospholipids

End of fatty acid side chains of
phospholipids

Cholesterol

Fig. 7.1   Chemical diagram of unit membrane and membrane as seen in
electron microscope. (From Loewy, A. G. and Siekevitz, P. 1969. *Cell Structure
and Function.* Holt, Rinehart and Winston.)

and that the types of contact between the surfaces vary. Such an observation is in line with observations from cell physiology about cell adhesion
to surfaces. It is, of course, particularly difficult to exclude fixation arte-
facts from electron microscopic studies of tissue cell interfaces, and this
explains the reluctance displayed by cytologists in making deductions
about cellular contact and adhesion from the electron microscope. However, it is now possible to say with confidence that electron microscopy
supports the idea of different kinds of cell to cell contact. The closest
contacts, termed tight junctions (Farquhar and Palade, 1963), appear to
involve complete contact or even fusion of the outer layers of two
adjacent cells. Epithelial tissue is particularly rich in tight junctions and
it is satisfying to note that, in electrophysiological experiments, many
adjacent epithelial cells show complete electrical continuity. It also seems
likely that small molecules pass easily between cells with such tight
junctions (Furshpan and Potter, 1968).

A second type of cell to cell contact may be termed a 'wide junction',
and it is the most frequent kind of surface association found between
adjacent tissue cells. The space between the cell surfaces is of the order of
10 to 20 nm wide. This space has posed problems for theorists wrestling
with the problem of cell adhesion, since most tightly adherent cells are
substantially linked by this type of contact. A popular explanation for
the gap is that it is actually filled with an intercellular 'cement', of low

electron density and therefore not discernible in the electron microscope. We will discuss the idea of intercellular cement a little later in the context of cell adhesion. An alternative view of wide junctions is that the gap is an artefact and that the cell surfaces are normally in much closer contact in these areas than the 10 to 20 nm observed. Certain cell junctions, which are much less common than these wide junctions and should not be confused with them, involve a gap of only about 2 nm between the cell surfaces, and are often termed 'gap junctions' (Gilula, Reeves and Steinbach, 1972).

The third kind of visible contact between tissue cells is of a highly specialized and localized nature, and seems to be characteristic of tightly adhesive cells. This contact appears as a plaque of dense material enveloping a short section of two cell surfaces, and is now known as a desmosome (Fig. 7.2). These desmosomes are not uncommon in adult epithelial tissue but are rare in embryonic tissues. Interestingly enough, desmosomes and tight junctions appear to be exclusive to vertebrate tissues, and other tissues display for the most part 'wide junctions', with a fourth type of structure, the septate desmosome occasionally present.

This brief review of the structure of cell junctions permits us to visualise the sorts of contacts between cells which are important in differentiation. A more comprehensive discussion of this topic will be found in Trinkaus (1959). How, then, do these cell to cell contacts affect the destiny of a cell? Probably the aspect in which cell contact is most crucially involved with differentiation is in embryonic induction. We have already briefly discussed the important observations that, in early embryogenesis, cell fate can be determined by cellular position and contact (see p. 38). Grobstein's (1961) work is particularly noteworthy as an attempt to study the nature of the cell contact in the process of induction. Using a system in which kidney tubule formation required induction by mesenchyme cells, he separated the two cell types by millipore filters of varying thicknesses and porosities. He was able to conclude that induction occurred without tight contact between cell surfaces, but that an 'intercellular matrix' played a crucial role in transmitting the morphogenetic information. This gives some support to the notion that 'wide junctions' as observed in the electron microscope are real, and are filled by non electron dense material. Further experiments on inductive interaction have been carried out by Saxen (1971) working on kidney development, and in these experiments trans-filter induction did not occur if cell contact was prevented.

Another experimental system which yields evidence about the passage of molecules between cells is the differentiating pancreas. Starting with epithelial tissue of pancreatic rudiment from embryonic mice, Rutter *et al.* (1968) have shown that contact with mesenchymal tissue is an essential prerequisite for the differentiation of the epithelial cells into pancreatic tissue. In this work, identification of pancreatic function depends on detection of amylase activity and zymogen granule concen-

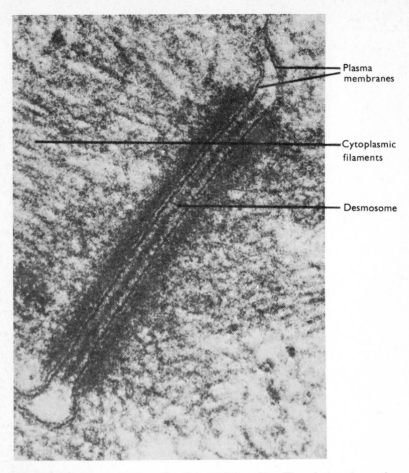

Plasma
membranes

Cytoplasmic
filaments

Desmosome

Fig. 7.2   A desmosome in cells of Amphibian skin. The picture shows the junction between two cells (× 210 000). (From Loewy, A. G. and Siekevitz, P. 1969. *Cell Structure and Function*. Holt, Rinehart and Winston.

tration. Pancreatic differentiation proceeds *in vitro* even when the epithelial and mesenchyme cell layers are separated by a millipore filter of 20 μm, and even when the source of mesenchyme tissue is very distant from the normal differentiating pancreas. These observations underline the fact that the differentiating substances exchanged between cells may often be widely distributed, and that they may travel over quite long distance. In other words, cell contact as a differentiating force need not involve close contact at all (see also Fig. 7.6).

Cell contact is, of course, necessary for cell adhesion, and cell adhesion is, itself, a potent factor in the process of differentiation. When experiments on mosaic and regulative development were discussed (see p. 38) it was apparent that cell movement from one embryonic site to another may often permit differentiation in a new direction. Conversely, if cell contact permits strong cellular adhesion, the fate of a cell will be closely bound up with its neighbouring cells and it will be exposed only to a confined spectrum of extracellular inducing molecules. An excellent statement of this aspect of cell adhesion comes from one of the pioneers in this field, Johannes Holtfreter '. . . the neural tube detaches itself from its ectodermal layer of origin, relinquishes also its previous tight adhesion to notochord and somites, and establishes new intimate contact relations to the outer ectoderm layer from which it had just departed. This newly achieved and stabilized anatomic constellation provides the opportunity for another set of inductions to occur' (Holtfreter, 1968). These implications of cell adhesion for differentiation clearly involve cell movement as well, and this topic will now be discussed more fully.

7.2 Cell movement

Cell movement and differentiation are interrelated in a variety of ways. Living cells move by cilia or flagella, by amoeboid movement, or by gliding over a solid substratum as do most animal tissue cells. We cannot discuss here the mechanics of such locomotory behaviour. It should, however, be emphasized that these types of locomotion are not only the results of differentiation but are, in many cases, an agent in initiating the differentiation process. Take, for example, the slime moulds already discussed in chapter 1. The motility of the amoeboid cells permits aggregation (Fig. 7.3), and the replacement of the individual movement of single amoebae by a coordinated movement of the entire slug permits the differential cell development of fruiting body and stalk (Fig. 7.4).

Many morphogenetic movements of cells and tissues are involved in differentiation and, especially in embryonic development, it appears that by moving from one site to another a cell might expose itself not simply to contact with other cells, but to what has come to be called 'positional information'. This phrase means, in brief, that a cell 'knows' where it is with reference to the whole organism. The problem has been explored closely in *Hydra* by Wolpert (1971). Because of its comparative simplicity and morphogenetic plasticity, *Hydra* is particularly suitable for grafting experiments and, by translocating cells from one part to another, Wolpert has been able to study the acquisition of positional information by the repositioned cells. His main conclusion is that gradients exist in the organism and sensitivity to these gradients supplies cells with information about where on the gradient they are located, so providing a clue to their overall position.

A particularly interesting but difficult problem arises in the interpretation of directed cell movements. It is that of choosing between differential adhesiveness, as against directed searching by the cell for a situation in which its positional information would be 'satisfied'. Whereas Steinberg has interpreted directed movement as a function of differential adhesiveness (Steinberg, 1970), Wolpert (1971) proposes a mechanism based on positional information. Whichever is the correct interpretation, and obviously both may apply in different organisms and situations, there is no doubt that cells not only become differentiated as a result of movement, but also move to particular positions as a result of their differentiation. Such a phenomenon has been termed 'sorting-out', and was elegantly displayed some years ago in experiments by Townes and Holtfreter (1955). Mixed aggregates of prospective epidermal and mesodermal cells, derived from an amphibian neurula, were cultered together. The initial randomly arranged mass of cells gradually transformed itself, by cell movement, into an organized form with epidermal cells on the outside and mesoderm inside (Fig. 7.5). Similar movements can be observed in mixed aggregates of chick embryonic tissue and it is certainly tempting to believe that such movements play some part in normal development.

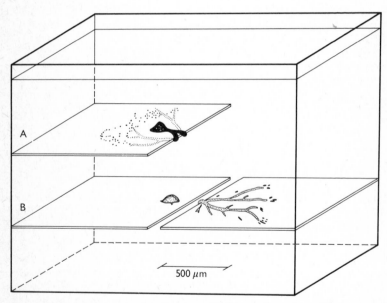

500 μm

Fig. 7.3 Aggregation in the slime mould *Dictyostelium*. A semidiagrammatic representation of two experiments done on aggregation in *Dictyostelium*, using coverslip shelves held under water. (A), the myxamoebae previously at random under the coverslip are attracted around the edge to the center on the upper surface; (B), the myxamoebae previously at random on the right hand coverslip are attracted to the centre on the left hand coverslip, across the substratum gap. In each case a preformed centre is acting as an attractant. (After Bonner, J. T. 1947. *J. exp. Zool.*, **106**, 1, 1–26. The Wistar Institute of Anatomy and Biology.)

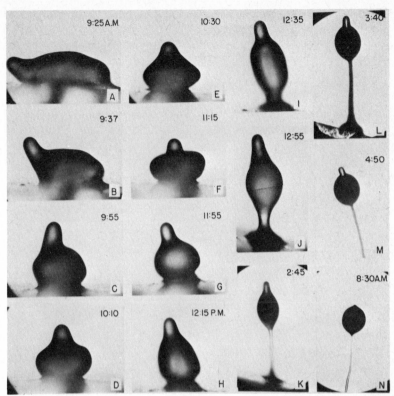

Fig. 7.4 Chronology of single developing sorocarp or fruiting body in *Dictyostelium discoideum*. (A) Migrating pseudoplasmodium. (B) End of migration. (C–D) Assumption of vertical orientation. (E–F) Initiation of stalk formation. (G–H) Beginning of sporophore elongation coincident with basal-disk formation. (I–J) Initiation of sorogen ascent. (K) Beginning of spore formation in peripheral area. (L–M) Inward progression during successive stages. (N) Mature sorus culmination completed previous evening. (From Berrill, N. J. 1961. *Growth, Development and Pattern*. W. H. Freeman & Co.)

A final comment on the role of cell movement in differentiation is that, especially in higher animals, some cells may move, or be moved, over very long distances in order to accomplish their differential destiny. Thus, a sperm, after long development in the testicular tissues and transport to the seminal vesicles, only utilizes its potential for flagellar movement after contact with the seminal fluids. An equally dramatic example is provided by the lymphoid cells of mammals. Although often carried passively in blood and lymph, they also move actively between other tissue cells (Fig. 7.6). Utilizing both mechanisms, the lymphoid cells which develop in the thymus during embryogenesis ultimately migrate to the spleen and lymph nodes and establish the future concentrations of lymphoid cells which occur in these organs (Gowans, 1966). Actually, even in adult life, the lymphoid cells appear to have both a travelling phase, during which they are transported in blood and lymph, and a resident phase during which they remain in one of the lymphoid organs.

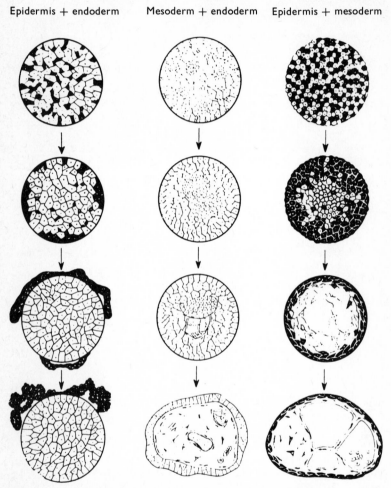

Epidermis + endoderm   Mesoderm + endoderm   Epidermis + mesoderm

Fig. 7.5   Cell sorting-out in embryonic tissues. Cells of the three germ layers are dissociated and mixed in various combinations. Combination of ectodermal and endodermal cells leads to self-isolation of the two tissues. Mesodermal and endodermal cells form a vesicle surrounded by endoderm and filled with derivatives of the mesodermal cells. Mesodermal and ectodermal cells form a vesicle surrounded by epidermis and filled with derivatives of the mesodermal cells. (From Townes, P. L. and Holtfreter, J. 1955. *J. exp. Zool.*, **128**, 1, 53–120. The Wistar Institute of Anatomy and Biology.)

7.3 Cellular response to surface molecules

The response of cells to hormone molecules has already been discussed in chapter 4, while their response to molecules carried on the surfaces of other cells has been mentioned earlier in this chapter. It is necessary, however, to consider another rather specialized cellular phenomenon— that of cell differentiation resulting from contact with molecules which are neither humoral nor cell borne. A well known example of such a phenomenon is the response of vertebrate leucocytes to phytohaemagglutinin. Phytohaemagglutinin (PHA) is a glycoprotein complex extracted from kidney beans, and its mitogenic (mitosis-stimulating) properties on human leucocytes determines their *in vitro* differentiation into plasma cells (Carstairs, 1961). A similar response by lymphocytes to tuberculin has been found in persons with a positive tuberculin reaction (Pearmain *et al.*, 1963). The mitogenic activity of PHA and tuberculin appears to be confined to lymphoid cells and it seems safe to conclude that this response by lymphoid cells may be analogous to the reaction of such cells to other antigens and that it is to be viewed, therefore, as a strictly immunological phenomenon with little or no relevance to the differentiation of other cells. This is not to minimize the intriguing aspect of differentiation involved in a clonal selection mechanism (see chapter 2),

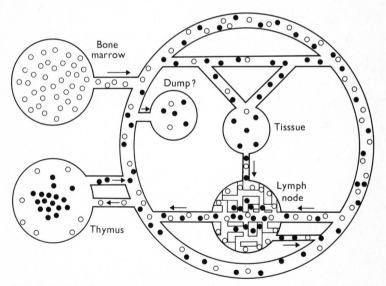

Fig. 7.6 Location and movements of lymphocytes. An idealization of the situations in mice in which lymphocytes are found and the primary links between them. Cells are to be thought of as thymus-derived (●), or bone-marrow-derived (○). The arrows indicate the usual direction of flow cells. (From Davies, A. J. S., Leuchers, E., Wallis, V. and Doenhoff, M. J. 1971. *Proc. R. Soc. Lond.*, B, **176**, 369–84, Fig. 1.)

but simply to recognize that the phenomenon is not of more general application.

Another case in which molecules external to the cell are known to be greatly involved with differentiation is the amoeba *Naegleria gruberi*. This organism will grow for many generations in the amoeboid form but, under certain culture conditions, will transform into a flagellated cell. These flagellates are temporary, non-reproductive and eventually revert to an amoeboid form (Willmer, 1958). Fulton (1972) has incubated *Naegleria* amoebae in a variety of media, and finds that two agents are effective in inducing the transformation from amoeba to flagellate— electrolytes and a low molecular weight factor present in yeast extract. Actually the two agents display a synergistic effect in the induction of transformation. There is an interesting link in the implication of ions in differentiation between transformation of *Naegleria* cells and the gene switching displayed by *Drosophila* giant chromosomes in Kroeger's experiments. He finds that the transcriptional activation of specific bands of these chromosomes can be induced by exposure to particular ions and ionic concentrations (Kroeger, 1966). I have elaborated elsewhere on the possible responses of the genes to altered ionic concentrations (Hilder and Maclean, 1974; Maclean and Hilder, 1976).

7.4 General comment

Our consideration of the cell surface in differentiation has been built round examples of animal cells only. The most significanct reasons for this are (a) that plants possess a rigid cell wall outside the membrane, reducing the possibilities for cell adhesion and (b) that plant tissues are relatively static and only in very early embryonic tissues does cell movement make any contribution to morphogenesis.

8

Controlled and uncontrolled differentiation

In the last chapter of this book we shall be looking at both artificially manipulated differentiation and pathologically redirected differentiation. The two topics are related, in that interference with the chromosome complement may itself both invoke a neoplastic change or be used to modify a malignant property already present.

8.1 Gene expression in allophenic mice

Allophenic animals are individuals which result from the mechanical combination of two blastocysts of differing genetic endowment. Cells with the differing genetic constitutions of the two donor blastocysts may then be detected in the adult by utilizing suitable genetic markers. The name of Dr. Beatrice Mintz has for some years been closely associated with the study of allophenic mice and her recent review provides a good account of this field (Mintz, 1971).

Mintz's method for the production of allophenic mice involves the collection of blastocysts from mice of differing genetic constitution. These are mechanically removed from donor females at about first cleavage and cultured in a medium with a high serum content. Following removal of the *zona pellucida* by pronase digestion, two embryos from different donors are placed in contact and incubated at 37 °C. After about 24 hours in culture many of these embryo pairs have fused to yield double-sized morulae of mixed origin which are now surgically implanted into the uterus of an 'incubator' female mouse, already made pseudopregnant by hormone injection. Soon after implantation the abnormally large embryo regulates to one of normal size. In Dr. Mintz's laboratory, approximately one-third of all such embryos survive to birth and their postnatal viability is very good (Mintz, 1965).

Mice of mixed genetic origin are, of course, only of use if derived from cells carrying good genetic markers. The markers used in Mintz's work are chiefly ones affecting coat colour and its distribution, but other antigenic and isozymic markers have also been investigated. In one group of 129 mice resulting from a C3H and C57 BL/6 genotypic combination, for example, 6 or more different tissues were analysed in each animal. Of the whole batch of mice, about 70% proved to be genetically mosaic.

No cell type has been found in which mosaicism can never be detected, although in any individual animal a particular tissue may prove to be exclusively of one genetic make-up. The frequency of mosaicism varies from one tissue to another however, and in the 90 mosaics out of the original 129 fused embryos, 40% had mosaic livers and 71% mosaic hair patterns (see Table 8.1).

Table 8.1   Clonal numbers in the mouse. (From Mintz, B. 1971. *Symp. Soc. exp. Biol.*, **25**, 345–70. Table 2.)

Cell type	No. of clones	Marker loci used
'Embryo' cells	3?	*Mdh-1, Gpi-1, Id-1*, etc.
Germ cells	2–9	*A, c*, etc.
Visual retina cells	20	*rd*
Melanoblasts	34	*B, c, d, ln, ru*, etc.
Hair follicle cells (mesodermal)	Approx. 170	*A, fz*

A clone is a population of cells derived from one original cell. Each tissue cell type can therefore be traced, by using suitable cell markers, to a basic number of original cells which are the original cells of that tissue. As the table shows, for the tissues so far studied the number of embryonic tissue original cells is always greater than one.

These experiments are particularly illuminating in terms of cell differentiation, since they indicate that individual groups of differentiated cells such as blood cells can be derived from more than one basal cell. Dr. Mintz has crystallized her deductions from these experiments into a clonal theory of tissue development, in which each tissue is derived from two or more clonal initiator cells. She defines a clone as 'the mitotic progeny of one cell in which a specific constellation of gene loci first became active (or derepressed) and has remained either active or mobilizable as a cell heredity'. In searching for an explanation for the multiclonal derivation of any one cell type, it is tempting to suppose that some protection against the expression of deleterious genes might be involved. In general, the precise mechanisms leading to multiclonal genetic expression are not known, and are probably diverse. X-chromosome inactivation, already discussed on page 67 is, of course, one such mechanism, and leads to the tortoiseshell coat colour in female mice and cats, which are heterozygous for the X-linked colour gene. But in allophenic animals the clonal patterns of gene expression involve genes not located on the X-chromosome and therefore not open to repression by persistent heterochromatinization. Dr. Mintz suggests that all clones of a cell type might have the same general batteries of genes active, but clonal differences in the expression of these genes might arise and become stabilized in a cell and its progeny. By virtue of their differing phenotypic properties, such clones would compete together in tissue development and might be affected by clonal selection.

The clonal distribution of coat colour genes in mice is particularly clearly demonstrated by allophenic animals. Such individuals, when mosaic for coat colour expression, display a lateral distribution of the clones so that differing colours tend to present as bands on the body (Fig. 8.1), each band presumably representing a clone derived from one cell. It is interesting to notice that many normal animals, not only mice, but dogs, cats and for example, tigers, tend to display a coat colour distribution in lateral bands. It must be emphasized that Mintz's experiments give no evidence for a general application of the allelic exclusion mechanisms or clonal selection theories of cells producing immuno-globulins (see p. 80) though evidence for such phenomena should be diligently sought in non-lymphoid tissues. We should be clear in our minds, moreover, that in many cases allelic exclusion is known *not* to occur. For example, individuals heterozygous for the allelic sickle cell form of β-globin, possess erythrocytes which will all sickle slightly at reduced oxygen tensions. If such individuals possessed erythrocytes of multiclonal origin, some expressing the normal β-globin allele and some the sickle form allele, then the blood cells would show a pronounced sickling dimorphism, some behaving like the erythrocytes of normal individuals and some like the erythrocytes of homozygous individuals with complete sickle cell anaemia (Schneider and Haggard, 1955). In other words, while gene expression in normal differentiation sometimes

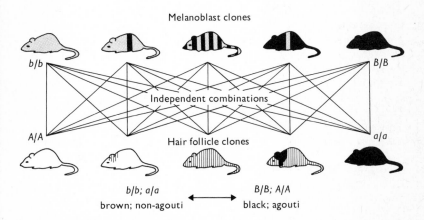

Fig. 8.1   Coat patterns in allophenic mice. Patterns realized in single allophenic animals as a result of combination of some of the possible independently originating melanoblast clonal patterns and hair follicle clonal patterns. The patterns demonstrate that the distribution of cells determining coat colour is clonal rather than random. (After Mintz, B. 1971. *Symp. Soc. exp. Biol.*, **25**, 345–70, Fig. 2.)

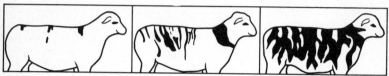

Fig. 8.2   Clonal patterns in skin pigment and hair follicles. (A) Melanoblast clones visible on the head of a human pigmentary mosaic. (B) Patterns of hair follicle clones in mosaic sheep. (After Mintz, B. 1971. *Symp. Soc. exp. Biol.*, **25**, 345–70, Figs. 3 and 4.)

falls roughly in line with the results observed in allophenic individuals (Fig. 8.2), often it does not.

8.2 Cancer and differentiation

The change to neoplasia and malignancy is a characteristic of most cell types of higher animals. Few species or tissues are exempt from the possibility of malignant change, although the liability does vary greatly between different species and different tissues. Cancers are comparatively rare in amphibians and frequent in man, uncommon in the small intestine but common in the stomach and the rectum: with the possible exception of crown gall, cancers do not occur in plants. So widespread is the incidence of cancer that it is justifiable to define it as a potential state of growth of animal cells. Cancer is often described as dedifferentiation, but this is only partially correct. Certainly most malignant cells have growth rates more characteristic of embryonic tissues, and often this growth rate itself involves changes in cellular morphology which could be styled dedifferentiation. On the other hand, myelomas, which are tumours of the lymphoid cells responsible for antibody production, continue to secrete globulins. Moreover, the globulin of each myeloma continues to be specific to the cell clone from which it arose (Saunders and Wilder, 1971). Rather than view the change to malignancy as dedifferentiation, it is more accurate to recognize it as a further phase of differentiation superimposed on an already differentiated cell. Therefore, although changing or modifying the original character of a cell somewhat, the basal differentiated commitment of the cell persists in the new malignant phase.

It is not possible to discuss here the many interesting aspects of cancer which impinge on differentiation, but it is appropriate to point out certain distinctions between normal differentiation and malignancy, as well as underlining certain similarities.

(a) Distinctions between malignancy and normal differentiation

1 Normal differentiation is sometimes reversible, while malignancy is very frequently so.
2 Malignant cells from a variety of tissues are more like one another than are the native tissues. Thus differentiated liver, skin or lymphoid cells are histologically and biochemically very different. However, their malignant derivates have common properties of invasive growth, reduced cellular adhesiveness and high glycolytic metabolism.
3 Differentiation is a tightly controlled process which follows certain predictable pathways, whereas malignancy is an accidental and relatively random event in the growth and development of a tissue or an organism.

To the extent that these generalizations are true, I would take issue with Bullough and Deol in their attempts to explain malignancy in terms of chalone activity, e.g. 'The basic difference between a tumour and its tissue of origin is not that its cells have changed their essential nature (though they may have lost certain of their original characteristics) but that the normal point of balance between cell gain and cell loss has shifted in favour of cell gain' and 'the basic difference between tumour cells and their tissue of origin is one of degree rather than of kind' (Bullough and Deol, 1971).

Conversely, many parallels between normal differentiation and carcinogenesis can be drawn (Dustin, 1972). For example:

1 Both normally differentiated and malignant cells are highly committed to specific metabolic and morphological characteristics and, in division, give rise to cells similarly committed. This memorizing of commitment is particularly striking both in tissue cultures of malignant cells and in, for example the cells of imaginal discs of fruit flies (Hadorn, 1965). Thus commitment to malignancy closely resembles commitment to other differentiating characters.
2 The notorious invasiveness of malignant cells is not necessarily an attribute of dedifferentiation, but may just as accurately be considered to be a further differentiating development of the malignant cell. Some non-malignant, but otherwise differentiated, tissues may also display this property, for example the embryonic trophoblast and, possibly, the regenerating nerve (Piatt, 1956).
3 Neither differentiation nor malignancy normally involve specific

morphological changes in chromosome number or form though exceptions are known, e.g. chromosome elimination in *Parascaris* (see p. 65) and the Philadelphia (Ph′) chromosome associated with chronic myeloid leukaemia in man (O'Riordan *et al.*, 1971).

(b) Malignancy and artificial heterokaryons

Although, at the present moment, the last statement regarding the absence of evidence for specific chromosomal changes in malignancy still holds true, some recent experiments with hybrid cells certainly implicate the chromosomes with the malignant character. The experiments have been carried out chiefly in the laboratory of Professor Henry Harris at Oxford, and consist of the artificial fusion of highly malignant cells with relatively non-malignant cells. When the hybrid cells were compared to the malignant cells by injection into susceptible mice, the hybrids proved to be many times less malignant (Wiener, Klein and Harris, 1973). In an earlier experiment it had been found that fusions between two distinct highly malignant cell lines were only poorly malignant (Harris, 1971). It is not easy to see how any hypothesis relating malignancy to an overall balance in chromosome number or group (as proposed by Hitosumachi, Rabinowitz and Sachs, 1971) can be squared with these findings. The most exciting part of this work by Harris and his colleagues is the observation that A9 cells, of initial low malignancy, are less potent in suppressing malignancy in the hybrid if they lose certain specific chromosomes. The search for the identity of these chromosomes has already begun, (Allerdice, *et al.*, 1973).

A recurrent difficulty in science is that theorizing is easiest, but least profitable, in a poorly understood situation. The fusion of malignant and non-malignant cells remains such a poorly understood experimental area. Nevertheless, it seems excusable to point out in a book of this type, that, both the 'acquisition' of malignancy without detectable loss of DNA and the suppression of such malignancy by DNA-coded cytoplasmic characters, are observations curiously reminiscent of the episomes of bacteria. For, with bacterial episomes, characters may be acquired by the bacterium through the integration of an episomic element. Conversely, if integration involves exchange with a segment of the host cell DNA, then that DNA segment might be irretrievably lost and the cell become permanently bereft of a specific item of genetic information.

(c) Mammalian embryonic tumours

Some remarkable tumours of mammalian embryonic cells are known, and they have proved extremely valuable experimental systems both in terms of our knowledge of differentiation in general, and of the malignant change. They are termed either teratomas or teratocarcinomas. Teratomas are benign growths which consist of a haphazard assortment of several different adult tissues. They occur either as products of ovary or testis; in the case of the ovary they almost certainly develop from a parthenogenetic activation of an ovum, while in the testis they are derived from one of the primordial germ cells. Teratocarcinomas are highly malignant tumours which, as well as producing the varied types of differentiated cell, continue to produce the relatively undifferentiated 'embryonal' carcinoma cells which are responsible for the spread of the tumour. The biology and experimental use of these tissues has been reviewed by Pierce (1967) and Damjanov and Solter (1974).

Certain recent discoveries about these abnormal growths have greatly increased their interest and usefulness as a research tool. They are

(i) the discovery of an inbred strain of mice, strain 129, which has a very high incidence of spontaneous testicular carcinoma (Stevens and Little, 1954)

(ii) teratomas may be induced by the introduction of I-celled mouse eggs into the testis of a male mouse (Stevens, 1970)

(iii) it has proved possible to establish teratocarcinoma cells in culture. Most of these cell lines, while remaining undifferentiated in culture, will produce the repertoire of differentiated cell types characteristic of teratomas, when injected into a new host (Kahan and Ephrussi, 1970).

(iv) a single 'embryonal' cell of a teratocarcinoma can give rise to a complete teratocarcinoma following implantation in a new host (Kleinsmith and Pierce, 1964)

(v) if mice are injected intraperitoneally with teratocarcinoma cells, they develop so-called 'embryoid bodies' in the ascites fluid. These bodies closely resemble the morula stage of a normal embryo and, if removed and grown in culture, will differentiate into the varied cell types of the mature tumour (Damjanov and Solter, 1974)

(vi) the immunological responsiveness of the host animal greatly influences the future differentiation and responsiveness of any introduced teratocarcinoma stem cells (Solter, 1975)

(vii) when differentiation takes place within teratomas or teratocarcinomas, it does so by the formation of groups of dividing cells which tend to differentiate together in a certain direction. In other words, differentiation occurs in nests of cells together, and therefore closely resembles that occurring in the normal mammalian embryo (Solter, 1975).

Besides simply remarking on the great value of these growths as an experimental differentiating system, we should also notice here that these neoplastic growths show every sign of being tumours of embryonic origin. This comment brings with it the corollary that other tumours are, by comparison, not embryonic in nature. Such a situation emphasises the folly of thinking that all tumours are examples of 'dedifferentiated' cells. It is also highly significant that the differentiated cells produced by teratocarcinomas are not themselves malignant, and that any 'genetic loss' theory of carcinogenesis must be difficult to square with the apparently successful differentiation into normal cells displayed by the stem cells of teratocarcinomas.

An interesting experiment has been attempted by Jami *et al.*, (1973) in which mouse fibroblasts and mouse teratocarcinoma stem cells were fused together, prior to implantation in a new host. The resulting tissue was invariably a fibrosarcoma-type tumour. This result implies that the fibroblast function, e.g. collagen synthesis, persisted in the hybrid cells, but that further differentiation was prevented. It was not absolutely clear in these experiments, however, whether gene loss had or had not occurred in the fused cells. Some of the hybrid cells apparently lacked some chromosomes present in one of the parent cells.

8.3 Oncogenic viruses and cancer

The last paragraph suggests that cancer may involve chromosomal elements which are reminiscent of bacterial episomes. There is evidence that, for at least some tumours, these elements may be viruses.

That a possible relationship may exist between neoplasms and viruses is not a new idea. As early as 1906 it was reported that human warts could be transmitted by cell free filtrates (Ciuffor, 1907), and in 1911 the pioneering work of Rous began with the successful transmission of

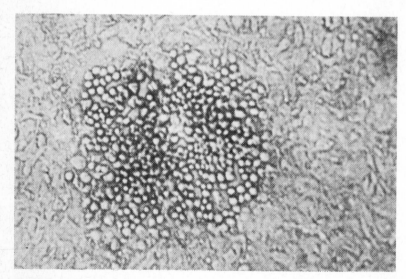

Fig. 8.3 Photograph of a group (focus) of chicken cells (magnification × 60) transformed by RSV infection. The spherical transformed cells are easily distinguished from the background of normal cells. (From Watson, J. D. 1970. *Molecular Biology of The Gene*. 2nd edition. Fig. 18.12, p. 618. W. A. Benjamin, Inc.)

malignant sarcoma in chicken by such filtrates. Some years elapsed, however, before the infective principle in these filtrates could be identified as definitely viral (Fig. 8.3). In the 1930's the Shope papilloma of rabbits and the Bittner milk factor inducing mouse mammary carcinoma were both suggested as indicative of a link between cancer and viruses (Shope and Hurst, 1933; Bittner, 1936). The discovery of polyoma virus by Gross (1970), and its capacity for tumour induction in a variety of rodents, has now permitted large scale investigation of cancer cell/oncogenic virus relationships.

A variety of different types of virus include particular members with oncogenic properties, and there are no grounds for thinking that the oncogenic viruses are a discrete and separate grouping. Many types of DNA virus include oncogenic examples e.g. pox virus, adenovirus, papovavirus, herpes virus (Table 8.2). Fewer RNA virus are known to be oncogenic (Table 8.3) and most of these belong to the large viruses known as leukoviruses.

The relationship between the virus and the cell is very different for the DNA and RNA oncogenic viruses. When a DNA virus is responsible for tumour induction a state of integration into the host chromosome is likely and production of free virus does not occur. Indeed, if viral production does occur the host cell may be destroyed, as in the normal cytolytic effect of most viruses. In studies with the papovavirus, polyma virus and SV 40, it is not only apparent that the alternative states of lytic viral multiplication or cell transformation to malignancy may occur with the same virus in different hosts, but also that, in one and the same tissue, some cells may be transformed and others lysed by the same virus

Table 8.2 Animal DNA viruses. (From Howatson, A. F. 1971. In *Comparative Virology*. (Eds.) K. Maramorosch and E. Kurstak. Table I, p. 512. Academic Press, Inc., New York)

Group	Mol. wt. of DNA ($\times 10^6$)	Morphology and dimensions (nm)	Envelope (+ or −)	Symmetry of capsid	Oncogenic members
Poxvirus	160–200	Brick-shaped 300 × 230	−	Complex	Rabbit fibroma, myxoma, Yaba molluscum contagiosum
Adenovirus	20–25	Spherical 70–80	−	Icosahedral	Adeno 12, 18, etc., in man. Simian and avian adenoviruses
Papovavirus	3–5	Spherical 45–55	−	Icosahedral	Polyoma, SV40, papilloma
Herpesvirus	60–80	Approx. spherical 100–150	+	Icosahedral	Lucke renal tumor Marek's disease Burkitt's lymphoma?
Parvovirus	2	Spherical 20	−	Icosahedral	Minute virus of mice?

(Howatson, 1971). Moreover, in some cases, the virus present in the transformed cell can be induced to enter a proliferative phase, as, for example, through the UV-irradiation of transformed cells (Gerber, 1964). Mammalian cells which have been transformed, by adenovirus, another type of DNA virus, have been found to include from 20–80 copies of the viral DNA and this DNA is the template for up to 5% of the cellular messenger RNA (Green, 1969).

The most striking difference in the RNA oncogenic viruses is the ability of the transformed cell to continue to produce infectious virus. In other words, the exclusive states of viral productivity or cellular transformation, applicable to DNA oncogenic viruses, do not hold for the RNA oncogenic viruses. Even so, the RNA viral-cell relations are often complex. The phenomena of viral interference, in which infection with one virus prevents or inhibits infection with virus of a different type, and helper virus, in which replication of one infecting virus is rendered possible only by superinfection with virus of a different type, have both been discovered in studies with these viruses. In addition some of the RNA viruses associated with leukaemia in a variety of mammals seem to have a very general distribution in the tissues of both leukaemic and normal animals. The last observation has led some virologists (Huebner and Todaro, 1969) to conclude that most vertebrate cells harbour RNA 'type C' viruses and that they may be normally transferred from parent to offspring (Figs 8.4 and 8.5). These authors argue that a variety of factors might be responsible for inducing an altered relationship between the virus and the host cell genome, and that one consequence of such an alteration might be a transformation to malignancy.

What then is the significance of cancer and oncogenic viruses in relation to cellular differentiation. Despite many years of intensive research,

Table 8.3   Animal RNA viruses. (From Howatson, A. F. 1971. In *Comparative Virology*. (Eds.) K. Maramorosch and E. Kurstak. Table II, p. 515. Academic Press, Inc., New York.)

Group	Mol. wt. of RNA ($\times 10^6$)	Morphology and dimensions (nm)	Envelope (+ or −)	Symmetry of capsid	Oncogenic members
Picornavirus	2–4	Spherical 20–30	−	Icosahedral	None known
Reovirus	10	Spherical 70–75	−	Icosahedral	Type 3?
Myxovirus	2–5	Approx. spherical 80–100	+	Helical	None known
Paramyxovirus	6–8	Pleomorphic 100–300	+	Helical	None known
Rhabdovirus	3–4	Bullet-shaped 70 × 170	+	Helical	None known
Leukovirus	10–12	Spherical 100–120	+	?	MLV, MSV, MTV, RSV, ALV, AMV, etc.

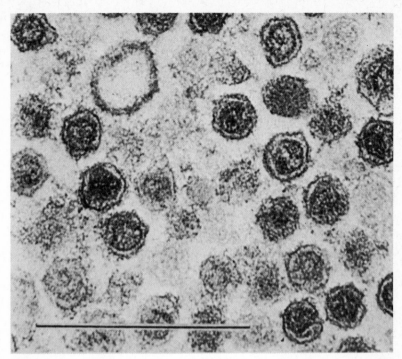

Fig. 8.4 Type C virus particles purified from leukemic mouse plasma, magnification, × 100 000. Bar represents 0.5 μm. (From Howatson, A. F. 1971. In *Comparative Virology*. (eds.) K. Maramorosch and E. Kurstak. Fig. 2. p. 516. Academic Press, Inc., New York.)

no entirely satisfactory general theory of carcinogenesis has emerged. We can still only construct a somewhat tentative hypothesis which will account for most of the available facts. Such facts as are known do suggest that particular genes on particular chromosomes are crucial in the change to malignancy, and that more than one such gene exists. Loss of these genes may render a cell malignant. Loss, in this context means effective loss of expression, which could result either from actual loss of the gene from the cell or some mechanical alteration preventing its proper transcription. A common mechanism for such gene loss seems to be the incorporation of a viral nucleic acid molecule, exchanged for the gene or implanted into it. Perhaps the loss of both copies on the two homologous chromosomes is normally required. It is appropriate to note that phage mu can integrate at any point in the chromosome of *E. coli* but, when it does so, the gene into which it is inserted is not expressed (Nomura and Engback, 1972). It is impossible to determine at the present time whether carcinogenic compounds and irradiations operate by inducing loss of such 'cancer' genes, or by affecting a virus which, in

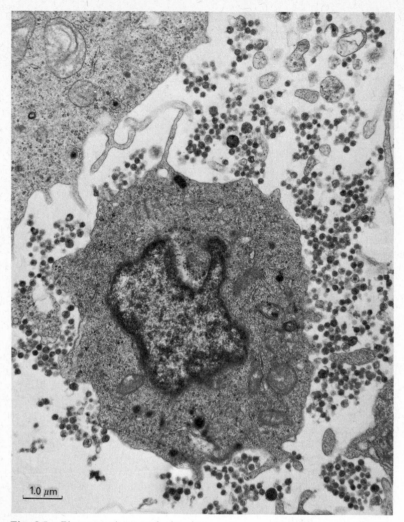

1.0 μm

Fig. 8.5   Electron micrograph showing a cat macrophage cell surrounded by C type virus particles. (From Oshiro, L. S. *et al.* 1971. *Cancer Res.*, **31**, 1100–10. Fig. 1.)

turn, leads to their loss. Surely the next decade will be a time during which the chromosomal location of such genes will be determined and the molecules for which they code identified.

Some implications of these ideas clearly have a direct bearing on differentiation. It is likely that viruses, in their varied and often persistent relations with eukaryotic cells, may be responsible for other differentiating characteristics besides malignancy. Some such possibilities have already been cited in chapter 6. Moreover, if accidental or programmed gene loss proved to be a common accompaniment of viral infection or even of normal cell aging, we should have to modify much of our original assumption that differentiated cells retain a complete genome. There is also evidence for gene transduction with animal viruses, so that tissue cells invaded by a virus may come to possess and express genes which previously belonged to another cell or, indeed, another species.

8.4   Chemical interference with differentiation—the use of BUdR

Many substances may be applied to cells or tissues which affect their differentiation, including mitotic inhibitors, hormones, antibiotics and carcinogens. Some of these we have already discussed in appropriate chapters. One substance does have some claim to longer and separate treatment, namely 5-Bromo-2′-deoxyuridine (BUdR). This substance is a nucleoside, an analog of thymidine, and is known to be incorporated into DNA in place of thymidine. It does act as a mutagen in some systems and its incorporation into bacterial and viral genomes leads to altered base pairing (Freese, 1959). More significantly, the compound has the ability to activate the genomes of viruses previously incorporated into bacterial and eukaryotic cell DNA, and to affect the cell growth and protein and nucleic acid synthesis of many eukaryotic cells. In many cases the effects of the compound appear to be highly selective and only the products of certain genes are affected. Thus its ability to alter the progress of many cells towards differentiation. Recent discussions of the effects of BUdR on differentiation will be found in Rutter, Pictet and Morris (1973) and Levitt and Dorfman (1974). Although the drug is toxic to some cells, many cell cultures which do show effects will return to normal growth if the compound is removed. There is even a subline of mouse melanoma cells where cell division is dependent on the drug (Davidson and Bick, 1973).

In the differentiating systems so far studied, BUdR appears to inhibit differentiation at some stage of development without markedly affecting the proliferation rate or viability of the cells. Its actions are not confined to the repression of particular proteins, since its presence in some tissues leads to increases in levels of alkaline phosphatase (discussed in Rutter, Pictet and Morris, 1973). These authors also draw attention to its apparent effects on some cell membranes, causing a tighter bond with the substrate, loss of malignant properties of a melanoma cell line, and the well established ability to activate viral genomes incorporated in cellular DNA. Table 8.4 lists the effects of BUdR observed in some differentiating systems.

At the moment, no single explanation for the diverse effects of BUdR is known. Indeed it seems likely that the effects arise by more than one mechanism. For example a cell surface effect seems likely as in the induction of dendrites in neuroblastoma cells, and in the altered adhesive

Table 8.4 Effects of Bromodeoxyuridine on Various Cells. Reproduced by permission. Rutter *et al.* (1973)

System	Parameter measured
Repression	
pancreatic acinar cells	exocrine enzymes
pancreatic B cells	insulin
chrondrocytes	chondroitin sulfate
myoblasts	myotube formation, myosin
pigmented retina cells	melanin
erythroblast precursors	hemoglobin
mammary gland	casein, α-lactalbumin
lymphocytes (primed)	antibodies
amnion cells	hyaluronic acid
liver cells (avian)	estrogen induction of phosvitin
hepatoma cells (HTC)	glucocorticoid induction of tyrosine aminotransferase
mouse lung	hyaluronic acid
mammary carcinoma hybrid cells	
melanoma	tumorigenicity, pigments
Induction	
mouse lung/mammary carinoma hybrid cells	alkaline phosphatase
mouse mammary carcinoma cells	alkaline phosphatase
pancreatic exocrine cells	alkaline phosphatase
neuroblastoma	neurite formation
	cell membrane glucoprotein
lymphoid cells (Burkitt lymphoma clones and NC37 line)	Epstein-Barr virus

properties of many cells. Substitution for thymidine seems to be a frequent explanation for some actions of the drug, particularly because its effects can often be mitigated or prevented by provision of excess thymidine. It is tempting to believe that the genes whose products are specifically affected might be those particularly rich in thymidine residues or dependent on T-rich regions for their control. Since DNA which has incorporated BUdR in place of thymidine is known to have a higher melting temperature (David, Gordon and Rutter, 1972, unpublished, quoted in Rutter, Pictet and Morris, 1973), these authors suggest that the drug might alter the dissociation constant of DNA-binding regulatory proteins. Such an effect might well present as repression or enhancement of specific protein synthesis.

BUdR effects on incorporated viral genomes are useful but puzzling. Since it seems to dislodge the appropriate segment from its colinear linkages in the cellular DNA, thus permitting cytoplasmic independence for the viral nucleic acid, it may be that this compound has the property of dissecting out other cellular genes, thus altering the genetic expression of the cell.

Whatever mechanisms emerge as the *modus operandi* of this compound, BUdR holds great promise as a tool in studies on cell differentiation.

9

Differentiation—The present perspective

Having reviewed evidence from many diverse biological systems which have a bearing on cellular differentiation, what can we conclude? What picture of differentiation can be squared with all, or even most of the evidence?

In my view, a differentiated cell is poised in a state of dynamic equilibrium between nucleus and cytoplasm, in which signals from the cytoplasm are necessary for the maintenance of the nuclear gene expression which characterizes that state. If nuclei are removed from differentiated cells and transplanted into foreign cytoplasm, the equilibrium is lost and the nucleus will then transcribe new genes. Moreover, the state of differentiation is not infinitely variable. It is insufficiently stressed in articles on differentiation that, even in the most complex multicellular organism, there is a strictly limited number of ultimate cell fates, with a maximum of a hundred or two.

Another striking fact about the differentiated state is that, although extremely stable, it is not absolutely so—take for example, the altered differentiation of cells in *Hydra* (Webster, 1971) and transdetermination in *Drosophila* (discussed in Postlethwait and Schneiderman, 1973). There seems little doubt that thoroughly differentiated cells of Hydra change the state of their differentiation to that of another cell type without an intervening mitosis. The message from transdetermination is slightly different since it involves cell division, and affects determined but not differentiated cells. In this process, imaginal disc cells, determined to a particular cell fate, do on rare occasions change their fate. The really interesting aspect of this is that the change is *directional*, so that a presumptive anal plate cell may become a presumptive leg, but very rarely *vice versa*. A pattern in fact exists, in which one determined state can be moved on to other determined states but, once moved, the cell behaves as if the new state were its original one. It, too, is stable.

These examples tell us that determined or overtly differentiated cells are in a state of equilibrium, but can be moved on to other destinies. However, the new destinies are not infinitely variable. If the stable state of determination or differentiation is broken, the cell may move to another, but the new state will be equally stable. In other words the nucleus and cytoplasm will have found a new equilibrium.

Imaginal discs also make plain to us how widely separated in time determination and differentiation may be. In other words that a pattern of gene expression may become inevitable long before the products of the genes are released.

The limited number of possible types of differentiated cell in a complex organism implies that a very limited number of 'tissue–master' genes may exist, each one responsible for the release of a particular genetic programme. In such a model the nucleo/cytoplasmic equilibrium need not directly involve all of the many genes active in any one cell. All that is needed is the control of the master genes and, given that, the gene expression characteristic of the cell type would follow as does the pre-set programme of a washing machine. It is not difficult to see how, in early development, the provision of a particular cytoplasmic environment by the subdivision of the egg will demand the activity of one master gene and the inactivity of the remainder. Put the nucleus in new cytoplasm and a new genetic programme may be expressed. Moreover, just as with the programming of a washing machine, it is sometimes possible to slide from one programme to another by tricking the machine, but not from *any* programme to *any* other, and only if the controlling elements are closely related. So with cells, one determined or differentiated state can, on rare occasions, be shifted on to another, but not to *any* other. The most frequent shifts will be those which are, so to speak, downhill.

We should notice here that this conception of differentiation provides an easy understanding of the distinction between determination and differentiation. The first is the assumption of the stable state between nucleus and cytoplasm which involves choice of one master gene, while the second is the actual expression of the programme under the direction of that gene. Clearly additional factors may often be necessary in order to release the activity of an already selected master gene. In addition, to describe each state as being stable is to imply not only that it is resistant to change within the life of that cell but also through many rounds of cell division. This clearly implies the existence of a strong cytoplasmic commitment to the stable state.

This discussion of 'tissue–master' genes and a stable nucleo-cytoplasmic equilibrium brings us back to a point made in the introduction to this book. Namely, that selective gene activity does not, of itself, explain differentiation. It is itself, in the view just outlined, merely the programmed expression of the activity of a 'tissue master' gene. The factors controlling and determining the activity or inactivity of tissue master genes are presumably often cytoplasmic, may well be partitioned during early embryonic cell division, and are the really decisive factors in cell differentiation. Their identity remains a mystery. Not only so, but these factors need not necessarily be molecular. Brief signals from the cellular environment relevant to cell position, cell movement, temperature or light might themselves be factors which could send a particular cell into the particular stable equilibrium in which a particular tissue master gene becomes, at least potentially, active. With such a philosophy the search for the identity of specific inducer molecules in early development could be a quest for the end of the rainbow.

This leads me to the question of what the future holds for biology in general and our knowledge and understanding of differentiation in particular. There is surely no doubt but that even a decade hence, experimentation in laboratories throughout the world will allow a much more precise understanding than is at present possible. Even the last five years have seen great progress, in that much more incisive and answerable questions are being asked about differentiation, and asking the correct questions is always half the battle. How are genes arranged on chromosomes, where is the genetic control machinery for eukaryotic gene expression, what does the heavy nuclear RNA consist of? All these, and many other questions are much more confined and precise questions than biologists were able to pose in the previous decade.

In particular the present experimental scene holds great promise for the future because of certain powerful techniques and systems which have recently come to the fore. Many times in the history of science, progress has been made in bursts, in rather sudden saltatory steps. This discontinuous advance is frequently due to the sudden flowering of ideas which had been growing steadily in many minds over long periods, for example Darwinian evolution by natural selection or the operon as a unit of genetic expression and control. But often the rapid surge in ideas and knowledge comes as a result of the exploitation of a new technique. Just as, in its day, the Feulgen reaction proved so informative about the location of DNA in cells and tissues, so also have the discovery of experimental systems such as phage infected bacteria, and anucleolate *Xenopus* mutants and the use of actinomycin D and high speed centrifugation been invaluable tools in biological research.

It is perhaps appropriate, then, to conclude this book by a short precis of the most important unanswered questions and the most promising new techniques in the general field of differentiation today. (The order in which questions and techniques are listed is of no significance.)

9.1 Questions of special significance in the field of differentiation

1 Are any cells, or any nuclei, beyond the point of a return to dedifferentiation?

2 Is it correct to assume that most differentiated tissue cells retain their full genetic complement?

3 What is the distribution of highly reiterated DNA in cells, and what are the factors governing the availability of these portions of DNA for transcription? Does the amount of this type of DNA vary during differentiation?

4 How much of differential gene expression is quantitative rather than qualitative? When it is said that a particular protein is not

synthesized in a cell, is this true down to the level of even a few molecules?

5 What is the mechanism of embryonic determination, and how do cells become fixed by, say, passing over the dorsal lip?

6 Are tissue cultures always aberrant tissues of their cell type, resulting from cell selection and/or viral involvement, or are they representative of cells in normal tissues?

7 What are the differences between cells in the same tissue in the young and the old animal or plant? Is cellular senescence real?

8 How valid and generally applicable is the chalone theory of control of organ size. If it is of only limited application, what controls the size of other cell populations, organs and tissues?

9 What is the nature of the cytoplasmic differences within the 'compartments' of a highly mosaic egg?

10 How is cell sorting out achieved and what is the role of differential adhesiveness in the process? Do cells release factors which affect their own adhesiveness and that of other cells?

9.2 Techniques and systems of special promise in the field of differentiation

1 Studies of mitochondrial and chloroplast form, function, and genetic material, both mutant and normal, are providing invaluable information about these relatively simple organelles and their relationship to the host cell. Such studies are, at the same time, very informative about cellular control machinery and its restriction or alteration during development and differentiation. It is a system in which the simplicity of a prokaryote yields information about the organization of a eukaryote.

2 *In vitro* hybridization, particularly of DNA to DNA using radioactive complementary DNA made via reverse transcriptase enzyme. Although it is likely that only multiple copy genes can be so located and the acquisition of bulk specific messenger RNA is necessary, the technique is a very powerful one.

3 The use of endonuclease restriction enzymes, which appear to be specific for palindromic sequences, promise a new lever on the question of gene arrangement on the DNA and the control of replication and transcription.

4 High voltage electron microscopy, especially with tilted pictures viewed in stereo, gives a new and better look at gross cellular morphology.

5 Cell fusion has already proved its worth in cell biology and it can surely be used further to investigate such problems as cellular senescence and nuclear dedifferentiation.

6 The banding pattern of mitotic chromosomes, as revealed by giemsa and quinacrine stains, and other histological procedures,

now provides an opportunity to identify individual chromosomes and therefore to push gene mapping further.

7 Induction of viral proliferation by treatment of cells with BUDR (Bromodeoxyuridine) and IUDR (iododeoxyuridine) is providing a very useful means of assaying for the presence of viral genomes in cells and will therefore provide more information on the effects of incorporated viruses on cells.

8 The exposure of differentiating cells to BUdR is also a most promising way of learning about the control of differentiation. An understanding of the mechanism of action of this drug, as discussed in chapter 8, may well provide, at the same time, an understanding of how the genetic control of differentiation is achieved.

9 Studies on teratomas and teratocarcinomas (discussed on pages 179 and 180) provide a most promising research system. These embryonic tumours constitute a situation in which apparently normal differentiated cells are derived from highly malignant embryonic stem cells. This point deserves careful scrutiny in the laboratory, probably by the implantation of the differentiated cell types into host animals. In addition, if the teratocarcinoma cells can be persuaded to differentiate or not, in culture, depending on culture conditions, then the factors important in the induction of cell determination and differentiation could well be assayed in such a controlled situation. Such a system has already been reported by Evans (1975).

What then of the future? Where are we in the understanding of differentiation, still at the beginning or within sight of the end? To my mind we are towards the end of the first stage, we now know how to grapple with some of the difficulties and we have some inkling about the size of the problem. But a glance at living cells under the microscope is still a sobering reminder of how little we understand.

References

ABELES, F. B. (1972). Action of ethylene. *A. Rev. Pl. Physiol.*, **23**, 259–72.

ALLARD, C., DE LAMIRANDE, G. and CANTERO, A. (1957). Behavior of enzymes in liver of starved rats. *Expl Cell Res.*, **13**, 69–77.

ALLERDICE, P. W., MILLER, O. J., MILLER, D. A., WARBURTON, D., PEARSON, P. L., KLEIN, G. and HARRIS, H. (1973). Chromosome analysis of two related heteroploid mouse cell livers by quinacrine fluorescence. *J. Cell Sci.*, **12**, 263–74.

ALLISON, A. C. (1959). Recent developments in the study of inherited anaemias. *Eugen. Q.* **6**, 155–64.

ARMS, K. (1968). Cytonucleoproteins in cleaving eggs of *Xenopus laevis*. *J. Embryol. exp. Morph.*, **20**, 367–74.

ASHWORTH, J. M. (1971). Cell development in the cellular slime mould *Dictyostelium discoideum*. In *Control Mechanisms of Growth and Differentiation. Symp. Soc. exp. Biol. no.* 25, 27–49.

ASHWORTH, J. M. (1973). *Cell Differentiation*. London: Chapman & Hall.

BARRETT, T., MARZANKA, D., HAMLYN, P. H. and GOULD, H. J. (1974). Non histone proteins control gene expression in reconstituted chromatin. *Proc. natn. Acad. Sci. U.S.A.*, **71**, 5057–61.

BASERGA, R. (1965). The relationship of the cell cycle to tumor growth and control of cell division: A Review. *Cancer Res.*, **25**, 581–95.

BEALE, G. H. (1969). A note on the inheritance of erythromycin-resistance in *Paramecium aurelia. Genet. Res.*, **14**, 341–2.

BECKER, H. J. (1956). *Verh. dt. zool. Ges.*, **14**, 341–2.

BECKER, H. J. (1966). Genetic and variegation mosaics in the eye of *Drosophila*. In *Current Topics in Developmental Biology*, Vol. 1. (ed. A. A. Moscona and A. Monroy), pp. 154–72. New York and London: Academic Press.

BECKWITH, J. R. and ZIPSER, D. (1970). *The Lactose Operon*. Cold Spring Harbor, N.Y.: Cold Spring Harbor Lab.

BELL, E. (1971). Information transfer between nucleus and cytoplasm during differentiation. In *Control Mechanisms of Growth and Differentiation. Symp. Soc. exp. Biol. no.* 25, 127–43.

BELL, P. R. (1970). Are plastids autonomous? In 'Control of Organelle Development'. *Symp. Soc. exp. Biol. no.* 24, 109–27.

BERENDES, H. D. (1971). Gene activation in dipteran polytene chromosomes. In *Control Mechanisms of Growth and Differentiation. Symp. Soc. exp. Biol. no.* 25, 145–61.

BERNHARD, W. (1964). Fine structural lesions induced by viruses. In *Ciba Found. Symp. Cellular Injury.* (ed. A. V. S. De Reuck and J. Knight), pp. 209–43. London: J. & A. Churchill, Ltd.

BERRILL, N. J. (1961). *Growth, Development and Pattern*. San Francisco and London: W. H. Freeman and Co.

BEUTLER, E., YEH, M. and FAIRBANKS, V. F. (1962). The normal human female as a mosaic of X-chromosome activity: Studies using the gene for G-6-PD-deficiency as a marker. *Proc. natn. Acad. Sci. U.S.A.*, **48**, 9–16.

BIRNSTIEL, M. L., CHIPCHASE, M. and SPEIRS, J. (1970). The ribosomal RNA cistrons. *Prog. Nucl. Acid Res.*, **11**, 351–89.

BIRNSTIEL, M., SPEIRS, J., PURDON, I. and JONES, K. Q. (1968). Properties and composition of the isolated ribosomal DNA satellite of *Xenopus laevis*. *Nature, Lond.*, **219**, 454–63.

BISHOP, J. O., PEMBERTON, R. and BAGLIONI, C. (1972). Reiteration frequency of haemoglobin genes in the duck. *Nature, Lond.*, (*New Biol.*), **235**, 231–44.

BITTNER, J. J. (1936). Some possible effects of nursing on the mammary gland tumor incidence in mice. *Science, N.Y.*, **84**, 162.

BLOOM, W. and FAWCETT, D. W. (1969). *A Textbook of Histology*. Philadelphia, London, Toronto: W. B. Saunders and Co.

BLUM, N., MALEKNIA, M. and SCHAPIRA, G. (1970). α- et β-globines libres et biosynthèse de l'hémoglobine. *Biochim. biophys. Acta*, **199**, 236–47.

BOLUND, L., RINGERTZ, N. R. and HARRIS, H. (1969). Changes in the cytochemical properties of erythrocyte nuclei reactivated by cell fusion. *J. Cell Sci.*, **4**, 71–87.

BONNER, J. T. (1947). Evidence for the formation of cell aggregates by chemotaxis in the development of the slime mold *Dictyostelium discoideum*. *J. exp. Zool.*, **106**, 1–26.

BONNER, J. (1965). *The Molecular Biology of Development*. Oxford: Clarendon Press.

BONNER, J. T. (1967). *The Cellular Slime Molds*. Princeton: Princeton Univ. Press.

BONNER, J. T., BARKLEY, D. S., HALL, E. M., KONIJN, E. M., MASON, J. W., O'KEEFE, G. and WOLFE, P. B. (1969). Acrasin, acrasinase, and the sensitivity to acrasin in *Dictyostelium discoideum*. *Devl Biol.*, **20**, 72–87.

BOSTOCK, C. J., PRESCOTT, D. M. and HATCH, F. T. (1972). Timing of replication of the satellite and main band DNAs in cells of the kangaroo rat. *Expl Cell Res.*, **74**, 487–95.

BOWERS, W. S., THOMPSON, M. J. and HEBEL, E. C. (1965). Juvenile and gonadotropic hormone activity of 10,11-epoxyfarneseric acid methyl ester. *Life Sci*, **4**, 2323–31.

BOYCOTT, A. E. and DIVER, C. (1923). On the inheritance of sinistrality in *Limnaea peregra*. *Proc. R. Soc.*, B. **95**, 207–13.

BRACHET, J. and FICQ, A. (1965). Binding sites of ^{14}C-actinomycin in amphibian oocytes and an autoradiography technique for the detection of cytoplasmic DNA. *Expl Cell Res.*, **38**, 153–9.

BREMNER, S., JACOB, F. and MESELSON, M. (1961). An unstable intermediate carrying information from genes to ribosomes for protein synthesis. *Nature, Lond.*, **190**, 576–81.

BRIEN, P. and RENIERS-DECOEN, M. (1950). Étude d'*Hydra viridis*. *Annls Soc. r. zool. Belg.*, **81**, 33–108.

BRITTEN, R. J. and DAVIDSON, E. H. (1969). Gene regulation for higher cells: a theory. *Science, N.Y.*, **165**, 349–57.

BRITTEN, R. J. and KOHNE, D. E. (1968). Repeated sequences in DNA. *Science, N.Y.*, **161**, 529–40.

BROWN, D. D. quoted in Flamm, W. G. (1972) pp. 36–7 (personal communication). Highly repetitive sequences of DNA in chromosomes. In *International Review of Cytology*, Vol. 32 (ed. G. H. Bourne and J. F. Danielli), pp. 1–51. New York: Academic Press.

BROWN, S. W. and NUR, V. (1964). Heterochromatic chromosomes in the coccids. *Science, N.Y.*, **145**, 130–6.

BUCHER, N. L. R., SCOTT, J. F. and AUT, J. C. (1951). *Cancer Res.*, **11**, 457–65. Regeneration of the liver in parabiotic rats.

BULLOUGH, W. S. (1967). *The Evolution of Differentiation*. London & New York: Academic Press.

BULLOUGH, W. S. and DEOL, J. V. R. (1971). The pattern of tumour growth. In *Control Mechanisms of Growth and Differentiation*. *Symp. Soc. exp. Biol. no.* 25, 255–75.

BURNET, F. M. (1959). *The clonal selection theory of acquired immunity.* Cambridge: Cambridge University Press.

BURNS, G. W. (1972). *The Science of Genetics. An Introduction to Heredity.* New York: Macmillan Co. London: Collier-Macmillan Ltd.

BUTCHER, R. W., ROBISON, G. A. and SUTHERLAND, E. W. (1972). Cyclic AMP and hormone action. In *Biochemical Actions of Hormones.* Vol. II (ed. G. Litwack), pp. 21–54. New York and London: Academic Press.

BUTLER, C. G. and SIMPSON, J. (1958). The source of the queen substance of the honey-bee. (*Apis mellifera* L.), *Proc. R. ent. Soc. Lond.*, Ser. A, **33**, 120–2

BYERS, T. J., PLATT, D. B. and GOLDSTEIN, L. (1963). The cytonucleoproteins of amoebae. II. Some aspects of cytonucleoprotein behavior and synthesis. *J. Cell Biol.*, **19**, 467–75.

CALLAN, H. G. (1963). The nature of lampbrush chromosomes. *Int. Rev. Cytol.*, **15**, 1–34.

CALLAN, H. G. (1967). The organization of genetic units in chromosomes. *J. Cell Sci.*, **2**, 1–7.

CALLAN, H. G. (1972). Replication of DNA in the chromosomes of eukaryotes. *Proc. R. Soc. Lond., B.*, **181**, 19–41.

CALLAN, H. G. (1973). Replication of DNA in eukaryotic chromosomes. *Brit. Med. Bull.* **29**, 192–5.

CAMPBELL, A. M. (1969). *Episomes.* New York, Evanston and London: Harper and Row.

CARLSON, J. G. (1952). Microdissection studies of the dividing neuroblast of the grasshopper *Chortophaga viridifasciata. Chromosoma*, **5**, 199–220.

CARSTAIRS, K. (1961). Transformation of the small lymphocytes in culture. *Lancet. ii.*, 984.

CHATTON, E. and LWOFF, A. (1935). Les ciliés apostomes. I. *Arch. Zool. exp. gén.*, **77**, 1–453.

CITRI, N. and POLLOCK, R. (1966). The biochemistry and function of β-lactanase (penicillinase). *Adv. Enzymol.*, **28**, 237–323.

CIUFFOR, G. (1907). Innesto positivo con filtrato di *Verruca valgare. G. ital. Mal. vener. Pelle*, **48**, 12–17.

CLAYTON, R. M. (1970). Problems of differentiation in the vertebrate lens. In *Current Topics in Developmental Biology*, vol. 5. (ed. A. A. Moscona and A. Monroy), pp. 115–80. New York and London: Academic Press.

CLEMENT, A. C. (1952). Experimental studies on germinal localization in *Ilyanassa. J. exp. Zool.*, **121**, 593–625.

CLOWES, R. C. (1972). Molecular structure of bacterial plasmids. *Bact. Rev.* **36**, 361–405.

COLE, R. J. (1975). In 'The early development of mammals'. *Br. Soc. Devl Biol. Symp.*, **2**, 335–57.

COLE, R. J. and PAUL, J. (1966). The effects of erythropoietin in haem synthesis in mouse yolk sac and cultured foetal liver cells. *J. Embryol. exp. Morph.*, **15**, 245–60.

COMFORT, A. (1955). In *Ciba Foundation Colloquia on Ageing*, vol. 1. 'General Aspects' (ed. G. E. W. Wolstenholme and M. P. Cameron), p. 244. London: J. & A. Churchill Ltd.

COMINGS, D. E. (1972). The structure and function of chromatin. *Adv. Human Genet.*, **3**, 237–431.

CONKLIN, E. G. (1905). The organization and cell lineage of the ascidian egg. *J. Acad. nat. Sci. Philad.*, **13**, 1.

COVE, D. J., PATEMAN, J. A. and REVER, B. M. (1964). Genetic control of nitrate reduction in *Aspergillus nidulans. Heredity, Lond.*, **19**, 529.

CRIPPA, M. and TOCCHINI-VALENTINI, G. P. (1971). Synthesis of amplified DNA that codes for ribosomal RNA. *Proc. natn. Acad. Sci. U.S.A.*, **68**, 2769–73.

CULLIS, C. A. (1973). DNA differences between flax genotrophs. *Nature, Lond.*, **243**, 515–16.

CUMMINS, J. E. (1969). The control of sequential DNA replication in *Physarum*. *Jap. J. Genet.*, **44**, Suppl. 1, 23–30.

CURTIS, H. J. and CROWLEY, C. (1963). Chromosome aberrations in liver cells in relation to ageing. In *Cellular Basis and Aetiology of Late Somatic Effects of Ionizing Radiation* (ed. R. J. B. Harris), pp. 251–8. London and New York: Academic Press.

DAMJANOV, I. and SOLTER, D. (1974). Experimental teratoma. *Cur. Top. Path.*, **59**, 69–131.

DAVIDSON, E. H. (1968). *Gene activity in early development*. London and New York. Academic Press.

DAVIDSON, R. L. and BICK, M. D. (1973). Bromodeoxyuridine dependence—a new mutation in mammalian cells. *Proc. natn. Acad. Sci. U.S.A.*, **70**, 138–42.

DAVIDSON, E. H., HASLETT, G. W., FINNEY, R. J., ALLFREY, V. G. and MIRSKY, A. E. (1965). Evidence for prelocalisation of cytoplasmic factors affecting gene activation in early embryogenesis. *Proc. natn. Acad. Sci. U.S.A.*, **54**, 696–704.

DAVIDSON, E. H., HOUGH, B. R., AMENSON, C. S. and BRITTEN, R. J. (1973). General interspersion of repetitive with non repetitive sequence elements in the DNA of *Xenopus. J. molec. Biol.*, **77**, 1–23.

DAVIES, A. J. S., LEVCHARS, E., WALLIS, V. and DOENHOFF, M. J. (1971). A system for lymphocytes in the mouse. *Proc. R. Soc. Lond. B*, **176**, 369–84.

DAVIS, D. G. and SHAW, M. W. (1964). An unusual human mosaic for skin pigmentation. *N. Engl. J. Med.*, **270**, 1384–9.

DAVISON, H. and DANIELLI, J. F. (1952). *Permeability of Natural Membranes*, 2nd edition. London: Cambridge University Press.

DAWID, I. B. (1972). Evolution of mitochondrial DNA sequences in *Xenopus. Devl Biol.*, **29**, 139–51.

DAWID, I. B. and BLACKLER, A. W. (1972). Maternal and cytoplasmic inheritance of mitochondrial DNA in *Xenopus. Devl Biol.*, **29**, 152–61.

DAYHOFF, M. O. (1972). *Atlas of Protein Sequence and Structure*, Vol. 5. Silver Spring, Md., U.S.A.: The National Biomedical Research Foundation.

DEE, J. (1962). Recombination in a Myxomycete, *Physarum polycephalum* Schw. *Genet. Res.* **3**, 11–23.

DE GOWIN, R. L., HOAK, J. C. and MILLER, S. H. (1972). Erythroblastic differentiation of stem cells in hemopoietic colonies. *Blood*, **40**, 881–92.

DE TERRA, N. (1970). Cytoplasmic control of macronuclear events in the cell cycle of *Stentor*. In *Control of Organelle Development. Symp. Soc. exp. Biol.*, no. 24, 345–68.

DONACHIE, W. D. and MASTERS, M. (1969). Temporal control of gene expression in bacterial. In *The Cell Cycle: Gene-Enzyme Interactions* (ed. G. M. Padilla, G. L. Whitson and I. L. Cameron), pp. 37–76. New York and London: Academic Press.

DOUGLAS, H. C. and HAWTHORNE, D. C. (1964). Enzymatic expression and genetic linkage of genes controlling galactose utilization in Saccharomyces. *Genetics, Princeton*, **49**, 837–44.

DU PRAW, E. J. (1970). *DNA and Chromosomes*. New York, London, Sydney: Holt, Rinehart and Winston, Inc.

DURRANT, A. (1962). The environmental induction of heritable change in *Linum. Heredity, Lond.*, **17**, 27–61.

DUSTIN, P. (1972). Cell differentiation and carcinogenesis: a critical review. *Cell & Tiss. Kinet.*, **5**, 519–33.

DWORKIN, M. (1973). Cell-cell interactions in the Myxobacteria. In *Microbial Differentiation. Soc. for Gen. Microbiol. Symp.*, **23**, 125–42.

ECHLIN, P. (1966). The cyanophytic origin of higher plant chloroplasts. *Br. phycol. Bull.*, **3**, 150–1.

ELGIN, S. C. R. and BONNER, J. (1970). Limited heterogeneity of the major non-histone chromosomal proteins. *Biochemistry*, N.Y., **9**, 4440–7.

ENAMI, M. (1951). The sources and activities of two chromatophorotropic hormones in crabs of the genus *Sesarma*. LI. Histology of incretory elements. *Biol. Bull. mar. biol. Lab., Wood's Hole*, **101**, 241–58.

Evans, M. (1975). Early mammalian development. Studies with pluripotent teratoma cells *in vitro*. In *2nd Symp. Soc. Devl Biol.*

FARQUHAR, M. G. and PALADE, G. E. (1963). Junctional complexes in various epithelia. *J. Cell Biol.*, **17**, 375–412.

FINCHAM, J. R. S. (1970). The regulation of gene mutation in plants. *Proc. R. Soc. Lond., B.*, **176**, 295–302.

FIRTEL, R. A. and LODISH, H. F. (1973). A small nuclear precursor of messenger RNA in the cellular slime mold *Dictyostelium discoideum*. *J. molec. Biol.*, **79**, 295–314.

FISCHBERG, M. and BLACKLER, A. W. (1963). Loss of nuclear potentiality in the soma versus preservation of nuclear potentiality in the germ line. In *Biological Organization at the Cellular and Supercellular level* (ed. R. J. C. Harris), pp. 111–27. London: Academic Press.

FISCHER, S., NAGEL, R. L. and FUHR, J. (1968). Interaction between reticulocyte polyribosomes and isolated human haemoglobin polypeptide chains. *Biochim. biophys. Acta*, **169**, 566–9.

FISHER, F. M. and SAMBORNE, R. C. (1964). *Nosema* as a source of juvenile hormone in parasitized insects. *Biol. Bull. mar. biol. Lab., Woods Hole*, **126**, 235–52.

FLAMM, W. G. (1972). Highly repetitive sequences of DNA in chromosomes. In *International Review of Cytology*, Vol. 32 (ed. G. H. Bourne and J. F. Danielli), pp. 1–51. New York: Academic Press.

FLAVELL, R. A. and JONES, I. G. (1971). DNA from isolated pellicles of *Tetrahymena*. *J. Cell Sci.*, **9**, 719–26.

FRAENKEL, G. (1935). A hormone causing pupation in the blowfly *Calliphora erythrocephala*. *Proc. R. Soc. Lond. B.*, **118**, 1–12.

FRASER, A. S. and SHORT, B. F. Hair follicle clonal patterns. (1960). *C.S.I.R.O. Res. Lab. Tech.* Paper no. 3. C.S.I.R.O. Melbourne, Australia.

FREESE, E. (1959). The specific mutagenic effect of base analogues on phage T_4. *J. molec. Biol.*, **1**, 87–105.

FRENSTER, J. H. (1965a). Nuclear polyanions as de-repressors of synthesis of ribonucleic acid. *Nature, Lond.*, **206**, 680–3.

FRENSTER, J. H. (1965b). A model of specific de-repression within interphase chromatin. *Nature, Lond.*, **206**, 1269–70.

FRENSTER, J. (1969). Biochemistry and molecular biophysics of heterochromatin and euchromatin. In *Handbook of Molecular Cytology*, ed. A. Lima-de-Faria. North Holland. Amsterdam.

FULTON, C. (1972). Early events of cell differentiation in *Naegleria gruberi*. Synergistic control by electrolytes and a factor from yeast extract. *Devl Biol.*, **28**, 603–19.

FURSHPAN, E. and POTTER, D. D. (1968). Low-resistance junctions between cells in embryos and tissue culture. In *Current Topics in Developmental Biology*, vol. 3 (ed. A. A. Moscona and A. Monroy), pp. 95–127. New York: Academic Press.

GALLY, J. A. and EDELMAN, G. (1972). The genetic control of immunoglobulin synthesis. *A. Rev. Genet.*, **6**, 1–45.

GALTSOFF, P. S. (1925). Regeneration after dissociation (an experimental study on sponges). I. Behavior of dissociated cells of *Microciona prolifera* under normal and altered conditions. *J. exp. Zool.*, **42**, 183–221.

GEORGIEV, G. P. (1969). On the structural organization of the operon and the regulation of RNA synthesis in animal cells. *J. theor. Biol.*, **25**, 473–90.

GERBER, P. (1964). Virogenic hamster tumor cells: induction of virus synthesis. *Science, N.Y.*, **145**, 833.

GHUYSEN, J. M. (1968). Use of bacteriolytic enzymes in the determination of wall structure and their role in cell metabolism. *Bact. Rev.*, **32**, 425–64.

GIBSON, I. (1970). The genetics of protozoan organelles. In *Control of Organelle Development. Symp. Soc. exp. Biol.* no. 24, 379–99.

GILBERT, W. and MULLER-HILL, R. (1966). Isolation of the *lac.* repressor. *Proc. natn. Acad. Sci. U.S.A.*, **56**, 1891–8.

GILULA, N. B., REEVES, O. R. and STEINBACH, A. (1972). Metabolic coupling, ionic coupling and cell contacts. *Nature, Lond.*, **235**, 262–5.

GOLDBERG, A. L. and DICE, J. F. *A. Rev. Biochem.*, **43**, 835–69 (1974). Intracellular protein degradation in mammalian and bacterial cells.

GOLDWASSER, E. (1966). Biochemical control of erythroid cell development. In *Current Topics in Developmental Biology*, vol. 1 (ed. A. A. Moscona and A. Monroy), pp. 173–211. New York: Academic Press.

GOWANS, J. L. (1966). Life, span, recirculation, and transformation of lymphocytes. *Int. Rev. exp. Path.*, **5**, 1–24.

GREEN, M. (1969). Messenger RNA in tumor cells induced by adenoviruses. In *Proceedings of 8th Canadian Cancer Research Conference*, **8** (ed. J. T. Morgan), pp. 261–85. London: Macmillan (Pergamon).

GREEN, M. M. (1969). Controlling element mediated transposition in the white gene in *Drosophila melanogaster. Genetics*, Princeton, **61**, 429–33.

GREGG, J. H. and BADMAN, W. S. (1970). Morphogenesis and ultrastructure in *Dictyostelium. Devl Biol.*, **22**, 96–111.

GROBSTEIN, C. (1961). Cell contact in relation to embryonic induction. *Expl. Cell Res. Suppl.*, **8**, 234–45.

GROS, F., HIATT, H., GILBERT, W., KURLAND, C. G., RISEBROUGH, R. W. and WATSON, J. D. (1961). Unstable ribonucleic acid revealed by pulse labelling of *Escherichia coli. Nature, Lond.*, **190**, 581–5.

GROSS, L. (1970). *Oncogenic Viruses*. Oxford: Pergamon Press.

GROSS, M. and GOLDWASSER, E. (1969). On the mechanism of erythropoietin-induced differention. V. Characterization of the ribonucleic acid formed as a result of erythropoietin action. *Biochemistry*, N.Y., **8**, 1795–805.

GROSSBACH, U. (1968). Cell differentiation in the salivary glands of *Camptochironomus tentans* and *C. pallidivittatus. Ann. zool. Fenn.*, **5**, 37–40.

GURDON, J. B. and UEHLINGER, V. (1966). 'Fertile' Intestine Nuclei. *Nature, Lond.*, **210**, 1240–1.

GURDON, J. B. and UEHLINGER, V. (1968). 'Fertile' intestine nuclei. In *Cellular Differentiation* (ed. J. R. Whittacher), pp. 2–8. Belmont, California: Dickerson Publishing Company, Inc.

GURDON, J. B. and WOODLAND, H. R. (1968). The cytoplasmic control of nuclear-activity in animal development. *Biol. Rev.*, **43**, 233–67.

GURDON, J. B. and WOODLAND, H. R. (1970). On the long-term control of nuclear activity during cell differentiation. In *Current Topics in Developmental Biology*, vol. 5 (ed. A. A. Moscona and A. Monroy), pp. 39–70. New York: Academic Press.

HADORN, E. (1965). Problems of determination and transdetermination. In *Brookhaven Symp. Biol.*, no. 18, pp. 148–61.

HADORN, E. (1967). Dynamics of determination. In *Major Problems in Developmental Biology* (ed. M. Locke), pp. 85–104. New York: Academic Press.

HALVORSON, H. O., CARTER, B. L. A. and TAURO, P. (1971). Synthesis of enzymes during the cell cycle. *Adv. microb. Physiol.*, **6**, 47–99.

HÄMMERLING, J. (1934). Regenerationsversuche an kernhaltigen und kernlosen Zellteilen von *Acetabularia Wellsteinii. Biol. Zbl.*, **54**, 650–7.

HÄMMERLING, J. (1953). Nucleo-cytoplasmic relationships in the development of Acetabularia. *Int. Rev. Cytol.*, **2**, 475–98.

HÄMMERLING, J. (1959). Spirogyra und Acetabularia (Ein vergleich ihrer fähigkeiten nach entfernung des kernes). *Biol. Zbl.*, **78**, 703–9.

HÄMMERLING, J. (1963). Nucleo-cytoplasmic interactions in *Acetabularia* and other cells. *A. Rev. Pl. Physiol.*, **14**, 65–92.

HARRIS, H. (1959). Turnover of nuclear and cytoplasmic ribonucleic acid in two types of animal cell, with some further observations on the nucleolus. *Biochem. J.*, **73**, 362–9.

HARRIS, H. (1970a). *Cell Fusion*. Clarendon Press.

HARRIS, H. (1970b). *Nucleus and Cytoplasm*. Oxford: Clarendon Press.

HARRIS, H. (1971). Cell fusion and the analysis of malignancy. *Proc. R. Soc. Lond. B*, **179**, 1–20.

HARRIS, H. and COOK, P. R. (1969). Synthesis of an enzyme determined by an erythrocyte nucleus in a hybrid cell. *J. Cell Sci.*, **5**, 121–33.

HARRIS, H. and WATKINS, J. F. (1965). Hybrid cells derived from mouse and man: artificial heterokaryons of mammalian cells from different species. *Nature, Lond.*, **205**, 640–6.

HENNIG, W. and WALKER, P. M. B. (1970). Variations in the DNA from two rodent families (Cricetidae and Muridae). *Nature, Lond.*, **225**, 915–19.

HERSKOWITZ, I. H. (1973). *Principles of Genetics*. London: Collier-Macmillan Ltd.

HILDER, V. A. and MACLEAN, N. (1974). Studies on the template activity of 'isolated' *Xenopus* erythrocyte nuclei, 1. The effects of ions. *J. Cell Sci.*, **16**, 133–42.

HITOSUMACHI, S., RABINOWITZ, Z. and SACHS, L. (1971). Chromosomal control of reversion in transformed cells. *Nature, Lond.*, **231**, 511–14.

HOLTFRETER, J. and HAMBURGER, V. (1955). Embryogenesis: Progressive differentiation. I. Amphibians. In *Analysis of Development* (ed. B. H. Willier, P. A. Weiss and V. Hamburger), pp. 230–96. Philadelphia: W. B. Saunders Co.

HOLTFRETER, J. (1968). Mesenchyme and epithelia in inductive and morpho-genetic processes. In *Epithelial-Mesenchymal Interactions*. 18th Hahnerman Symposium (ed. R. Fleischmajer and R. E. Billingham), pp. 1–30. Baltimore: The Williams and Wilkins Company.

HOWATSON, A. F. (1971). Oncogenic viruses: a survey of their properties. In *Comparative Virology* (ed. K. Maramorosch and E. Kurstak), pp. 509–37. New York and London: Academic Press.

HUANG, R. C. C. and HUANG, P. C. (1969). Effect of protein-bound RNA associated with chick embryo chromatin on template specificity of the chromatin. *J. molec. Biol.*, **39**, 365–78.

HUBER, R. and HOPPE, W. (1965). Zur chemie des ecdysons, VII. Die kristall-und molekülstrukturanalyse des insektenverpuppungshormons ecdyson mit der automatisierten faltmolekulmethode. *Chem. Ber.*, **98**, 2403–24.

HUBERMAN, J. A. and RIGGS, A. D. (1968). On the mechanism of DNA replication in mammalian chromosomes. *J. molec. Biol.*, **32**, 327–41.

HUEBNER, R. J. and TODARO, G. J. (1969). Oncogenes of RNA tumor viruses as determinants of cancer. *Proc. natn. Acad. Sci. U.S.A.*, **64**, 1087–94.

HYMAN, L. (1940). *The Invertebrates, Protozoa through Ctenophora*. New York: McGraw-Hill.

INGRAM, V. M. (1961a). Gene evolution and the haemoglobins. *Nature, Lond.*, **189**, 704–8.

INGRAM, V. M. (1961b). *Hemoglobin and its Abnormalities*. Springfield, Illinois: C. C. Thomas.

ITANO, H. A. (1957). The human hemoglobins: their properties and genetic control. *Adv. Protein Chem.*, **12**, 215–68.

JACOB, F. (1968). Regulatory devices in the bacterial cell. In *Cellular Differentiation* (ed. J. R. Whittaker), pp. 34–7. Belmont, California: Dickenson Publishing Company, Inc.

JACOB, F. and MONOD, J. (1961). Genetic regulatory mechanisms in the synthesis of proteins. *J. molec. Biol.*, **3**, 318–56.

JAMI, J., FAILLY, C. and RITZ, E. (1973). *Expl Cell Res.*, **76**, 191–9. Lack of expression of differentiation in teratoma-fibroblast somatic cell hybrids.

JEON, K. W. and LORCH, I. J. (1969). Lethal effect of heterologous nuclei in amoeba heterokaryons. *Expl. Cell Res.*, **56**, 233–8.

JINKS, J. L. (1964). *Extrachromosomal Inheritance*. Prentice-Hall foundations of modern Genetics series. Englewood Cliffs, New Jersey: Prentice-Hall.

JOHNSON, R. T. and HARRIS, H. (1969). DNA synthesis and mitosis in fused cells. *J. Cell Sci.*, **5**. I. HeLa homokaryons, 603–24. II. HeLa chick erythrocyte heterokaryons, 625–43. III. HeLa Ehrlich heterokaryons, 645–97.

JOHNSON, R. T. and RAO, P. N. (1971). Nucleo-cytoplasmic interactions in the achievement of nuclear synchrony in DNA synthesis and mitosis in multinucleate cells. *Biol. Rev.*, **46**, 97–155.

JONES, K. W. (1970). Chromosomal and nuclear location of mouse satellite DNA in individual cells. *Nature, Lond.*, **225**, 912–15.

JOHNSON, R. T. and RAO, P. N. (1970). Mammalian Cell Fusion: Induction of Premature Chromosome Condensation in Interphase Nuclei. *Nature, Lond.*, **226**, 717–22.

JUKES, T. H. (1966). *Molecules and Evolution*. New York and London, Columbia University Press.

JUKES, T. H. and HOLMQUIST, R. (1972). Estimation of evolutionary changes in certain homologous polypeptide chains. *J. molec. Biol.*, **64**, 163–79.

JURAND, A. and SELMAN, G. G. (1969). *The anatomy of Paramecium aurelia*. London: Macmillan.

KAHAN, B. W. and EPHRUSSI, B. (1970). Developmental potentialities of clonal *in vitro* cultures of mouse testicular teratoma. *J. natn. Cancer Inst.*, **44**, 1015–29.

KALMUS, H., KERRIDGE, J. and TATTERSFIELD, F. (1954). Occurrence of susceptibility to carbon dioxide in *Drosophilia melanogaster* from different countries. *Nature, Lond.*, **173**, 1101–2.

KEDES, L. H. and BIRNSTIEL, M. L. (1971). Reiteration and clustering of DNA sequences complementary to histone messenger RNA. *Nature, Lond., New Biol.*, **230**, 165–9.

KENDREW, J. C., WATSON, H. C., STRANDBERG, B. E., DICKERSON, R. E., PHILLIPS, D. C. and SHORE, V. C. (1961). The amino-acid sequence of sperm whale myoglobin. A partial determination by X-ray methods, and its correlation with chemical data. *Nature, Lond.*, **190**, 666–70.

KEYL, H. G. (1965). Duplikationen von Untereinheiten der chromosomalen DNS während der Evolution von *Chironomus thummi Chromosoma*, **17**, 139–80.

KLEINSMITH, L. J. and PIERCE, G. B. (1964). *Cancer Res.*, **24**, 1544–51. Multipotentiality of single embryonal carcinoma cells.

KNUTSEN, G. (1965). Induction of nitrite reductase in synchronised cultures of *Chlorella pyrenoidosa*. *Biochem. biophys. Acta*, **103**, 495–502.

KOSLOV, Y. V. and GEORGIEV, G. P. (1970). Mechanism of inhibitory action of histones on DNA template activity *in vitro*. *Nature, Lond.*, **228**, 245–7.

KREBS, E. G., HUSTON, R. B. and HUNKELER, F. L. (1968). Properties of phosphorylase kinase and its control in skeletal muscle. *Adv. Enzyme Regul.*, **6**, 245–55.

KROEGER, H. (1966). Potentialdifferenz und puff-muster. Elektrophysiolofische und cytologische untersuchunger an der speicheldrusen von *Chironomus thummi*. *Expl. Cell Res.*, **41**, 64–80.

KRUEGER, R. G. and MCCARTHY, B. J. (1970). Hybridisation studies with nucleic acids from murine plasma cell tumors. *Biochem. biophys. Res. Commun.*, **41**, 944–51.

KUHN, A. (1971). *Lectures on Developmental Physiology*. Berlin, Heidelberg, New York. Springer-Verlag.

KUNKEL, H. C. and WALLENIUS, C. (1955). New hemoglobin in normal adult blood. *Science, N.Y.*, **122**, 288.

LAJTHA, L. G. (1964). Recent studies in erythroid differentiation and pro-liferation. *Medicine, Baltimore*, **43**, 625–33.

LANE, C. D., MARBAIX, G. and GURDON, J. B. (1971). Rabbit haemoglobin synthesis in frog cells: the translation of reticulocyte 9S RNA in frog oocytes. *J. molec. Biol.*, **61**, 73–91.

LANGAN, T. A. (1968). Histone phosphorylation; stimulation by adenosine 3′,5′-monophosphate. *Science, N.Y.*, **162**, 579–80.

LASH, J. W. (1968). Chondrogenesis: Genotypic and phenotypic expression. *J. Cell Physiol.*, **72**, Suppl. 1, 35–46.

LEAK, L. V. (1967). Studies on the preservation and organization of DNA-containing regions in a blue-green alga, a cytochemical and ultrastructural study. *J. Ultrastruct. Res.*, **20**, 190–205.

LEAKE, R. E., TRENCH, M. E. and BARRY, J. M. (1972). Effect of cations on the condensation of hen erythrocyte nuclei and its relation to gene activation. *Expl. Cell Res.*, **71**, 17–26.

LEE, J. C. and INGRAM, V. M. (1967). Erythrocyte transfer RNA: change during chick development. *Science, N.Y.*, **158**, 1330–2.

LEHMANN, H. and CARRELL, R. W. (1969). Variations in the structure of human haemoglobin with particular reference to the unstable haemoglobins. *Br. med. Bull.*, **25**, 14–23.

LEVINE, M. (1969). Phage morphogenesis. *A. Rev. Genet.*, **3**, 323–42.

LEVITT, D. and DORFMAN, A. (1974). Concepts and mechanisms of cartilage differentiation. *Curr. Top. Devl. Biol.*, **8**, 103–49.

LEWIS, E. B. (1950). The phenomenon of position effect. *Adv. Genet.*, **3**, 73–115.

LEWIS, E. B. (1964). Genetic control and regulation of developmental pathways. In *The role of chromosomes in development* (ed. M. Locke), 231–52. New York and London: Academic Press.

LEWIS, K. R. and JOHN, B. (1968). The chromosomal basis of sex determination. *Int. Rev. Cytol.*, **23**, 277–379.

LINDBERG, V. and DARNELL, J. E. (1970). SV40-specific RNA in the nucleus and polyribosomes of transformed cells. *Proc. natn. Acad. Sci. U.S.A.*, **65**, 1089–96.

LISOWSKA-BERNSTEIN, B., LAMM, M. E. and WASSALLI, P. (1970). Synthesis of immunoglobulin heavy and light chains by the free ribosomes of a mouse plasma cell tumour. *Proc. natn. Acad. Sci. U.S.A.*, **66**, 425–32.

LODISH, H. F. (1970). Specificity in bacterial protein synthesis: role of initiation factors and ribosomal subunits. *Nature, Lond.*, **226**, 705–12.

LOEWY, A. G. and SIEKEVITZ, P. (1969). *Cell Structure and Function*. London and New York: Holt, Rinehart and Winston.

MACLEAN, N., HILDER, V. and BAYNES, Y. A. (1973). RNA synthesis in Xenopus enythrocytes. *Cell Differentiation*, **2**, 261–9.

MACLEAN, N. and JURD, R. D. (1971). The haemoglobins of healthy and anaemic *Xenopus laevis*. *J. Cell Sci.*, **9**, 509–28.

MACLEAN, N. and JURD, R. D. (1972). The control of haemoglobin synthesis. *Biol. Rev.*, **47**, 393–437.

MADGWICK, W. J., MACLEAN, N. and BAYNES, Y. A. (1972). RNA synthesis in chicken erythrocytes. *Nature, Lond. (New Biol.)*, **238**, 137–9.

MANES, C. and SHARMA, O. K. (1973). Hypermethylated and RNA in clearing rabbit embryos. *Nature, Lond.*, **244**, 283–4.

MARAMATSU, M., HODNETT, J. L., STEELE, W. J. and BUSCH, H. (1966). Synthesis of 28-S RNA in the nucleus. *Biochim. biophys. Acta*, **123**, 116–25.

MARGOLIASH, E., FITCH, W. M. and DICKERSON, R. E. (1971). Molecular expression of evolutionary phenomena in the primary and tertiary structures of cyto-chrome C. In *Molecular Evolution*, Vol. II, 'Biochemical Evolution and the Origin of Life' (ed. E. Schoffeniels), 52–95. Amsterdam: North-Holland Publishing Co.

MARKERT, C. L. and URSPRUNG, H. (1962). The ontogeny of isozyme patterns of lactate dehydrogenase in the mouse. *Devl Biol.*, 5, 363–81.

MARKERT, C. L. and URSPRUNG, H. (1971). *Developmental Genetics*. Foundations of developmental series. Englewood Cliffs, New Jersey. Prentice-Hall, Inc.

MARKS, P. A. and KOVACH, J. S. (1966). Development of mammalian erythroid cells. In *Current Topics in Developmental Biology*, Vol. 1 (ed. A. A. Moscona and A. Monroy), 213–52. New York: Academic Press.

MARKS, P. A. and RIFKIND, R. A. (1972). Protein synthesis: its control in erythropoiesis. *Science, N.Y.*, 175, 955–61.

McCLINTOCK, B. (1951). Chromosome organization and genic expression. *Cold Spring Harb. Symp. quant. Biol.*, 16, 13–47.

McCULLOCH, E. A. and TILL, J. E. (1962). The sensitivity of cells from normal mouse bone marrow to gamma radiation *in vitro* and *in vivo*. *Radiation Res.*, 16, 822–32.

MEDAWAR, P. B. (1957). *The uniqueness of the individual*. London: Methuen.

MILLER, C. O. (1970). Plant hormones. In *Biochemical actions of Hormones*, Vol. 1 (ed. G. Littwark). London and New York: Academic Press.

MILLER, O. L. and BEATTY, B. R. (1969). Portrait of a gene. *J. Cell Physiol.*, 74, Suppl. I, Part II, 225–32.

MILLER, O. L. and HAMKALO, B. A. (1972). Visualization of RNA synthesis on chromosomes. *Int. Rev. Cytol.*, 33, 1–25.

MINTZ, B. (1965). Genetic mosaicism in adult mice of quadriparental lineage. *Science, N.Y.*, 148, 1232–3.

MINTZ, B. (1971). Clonal basis of mammalian differentiation. In *Control Mechanisms of Growth and Differentiation, Symp. Soc. exp. Biol.*, no. 25, 345–70.

MITCHISON, J. M. (1971). *The Biology of the Cell Cycle*. Cambridge, Cambridge University Press.

MITCHISON, J. M. and CREANOR, J. (1969). Linear synthesis of sucrase and phosphatases during the cell cycle of *Schizosaccharomyces pombe*. *J. Cell Sci.*, 5, 373–91.

MONOD, J. and COHEN-BAZIRE, G. (1953). The specific inhibitory effect of β-galactisides in the biosynthesis (constitutive) of β-galactosidase in *Escherichia coli. C. r. hebd. Séanc. Acad. Sci., Paris*, 236, 417–19.

MORGAN, T. H. (1926). Genetics and the physiology of development. *Am. Nat.*, 60, 489–515.

NANNEY, D. L. (1966). Corticotype transmission in *Tetrahymena. Genetics, Princeton*, 54, 955.

NAORA, H. and KODAIRA, K. (1970). Interaction of informational macromolecules with ribosomes. II Binding of tissue-specific RNA's by ribosomes. *Biochim. biophys. Acta*, 209, 196–206.

NOLTE, D. J., MAY, I. R. and THOMAS, B. M. (1970). The gregarisation phenomena of locusts. *Chromosoma*, 29, 462–73.

NOMURA, M. and ENGBAEK, F. (1972). Expression of ribosomal protein genes as analyzed by bacteriophage Mu-induced mutations. *Proc. natn. Acad. Sci. U.S.A.*, 69, 1526–30.

NOSSAL, G. J. V. and LEDERBERG, J. (1958). Antibody production by single cells. *Nature, Lond.*, 181, 1419–20.

NOVICK, R. P. (1969). Extachromosomal inheritance in bacteria. *Bact. Rev.*, 33, 210–62.

O'BRIEN, B. R. A. (1961). Identification of haemoglobin by its catalase reaction with peroxide and O-dianisidine. *Stain. Technol.*, 36, 57–61.

OHNO, S. (1970). *Evolution by Gene Duplication*. London: Allen and Unwin.

O'MALLEY, B. W., ROSENFELD, G. C., COMSTOCK, J. P. and MEANS, A. R. (1972). Steroid hormone induction of a specific translatable messenger RNA. *Nature, Lond. (New Biol.)*, 240, 45–7.

ORD, M. J. (1968). The viability of the nucleate cytoplasm of *Amoeba proteus*. *J. Cell Sci.*, **3**, 81–8.

ORD, M. J. (1971). The initiation, maintenance and termination of DNA synthesis: A study of nuclear DNA replication using *Amoeba proteus* as a cell model. *J. Cell Sci.*, **9**, 1–21.

ORD, M. J. (1973). Changes in nuclear and cytoplasmic activity during the cell cycle with special reference to RNA. In *The Cell Cycle in Development and Differentiation*. *Br. Soc. for Develop. Biol. Symp.* (ed. M. Balls and F. S. Billett), 31–49. Cambridge: University Press.

O'RIORDAN, M. L., ROBINSON, J. A., BUCKTON, K. E. and EVANS, H. J. (1971). Distinguishing between the chromosomes involved in Down's syndrome (trisomy 21) and chronic myeloid leukaemia (Ph′) by fluorescence. *Nature, Lond.*, **230**, 167–8.

OSHIRO, L. S., RIGGS, J. L., TAYLOR, D. O. N., LENNETTE, E. H. and HUEBNER, R. J. (1971). Feritin-labelled antibody studies of feline c-type particles. *Cancer Res.*, **31**, 1100–10.

PARDUE, M. L. and GALL, J. G. (1970). Chromosomal localization of mouse satellite DNA. *Science, N.Y.*, 1356–8.

PAUL, J. (1971). Transcriptional regulation in mammalian chromosomes. In 'Control Mechanisms of Growth and Differentiation'. *Symp. Soc. exp. Biol.*, no. 25, 117–26.

PEARMAIN, G., LYCETTE, R. R. and FITZGERALD, P. H. (1963). Tuberculin-induced mitosis in peripheral blood leucocytes. *Lancet*, 637–8.

PETERKOFSKY, B. and TOMKINS, G. M. (1967). Effect of inhibitors of nucleic acid synthesis on steroid-mediated induction of tyrosine aminotransferase in hepatoma cell cultives. *J. molec. Biol.*, **30**, 49–61.

PFEIFFER, S. E. and TOLMACH, L. J. (1968). RNA synthesis in synchronously growing populations of HeLa S_3 cells. I. Rate of total RNA synthesis and its relationship to DNA synthesis. *J. Cell Physiol.*, **71**, 77–94.

PIATT, J. (1956). Studies on the problem of nerve pattern. I. Transplantation of the forelimb primordium to ectopic sites in *Ambystoma*. *J. exp. Zool.*, **131**, 173–201.

PIERCE, G. B. (1967). Teratocarcinoma: model for a developmental concept of cancer. *Curr. Top. Devl. Biol.*, **2**, 223–46.

PITELKA, D. R. (1969). Centriole replication. In *Handbook of Molecular cytology, North-Holland Research Monographs. Frontiers of Biology*, Vol. 15 (ed. A. Lima-de-Faria), 1199–218. Amsterdam: North-Holland Publishing Company.

POODRY, C. A., BRYANT, P. J. and SCHNEIDERMAN, H. A. (1971). The mechanism of pattern reconstruction by dissociated imaginal discs of *Drosophila melanogaster*. *Devl Biol.*, **26**, 464–77.

POSTLETHWAIT, J. H. and SCHNEIDERMAN, H. A. (1973). Developmental genetics of *Drosophila* imaginal discs. *A. Rev. Genet.*, **7**, 381–433.

PREER, J. R. (1971). Extrachromosomal inheritance: hereditary symbionts, mitochondria, chloroplasts. *A. Rev. Genet.*, **5**, 361–406.

RAO, P. N. and JOHNSON, R. T. (1970). Mammalian cell fusion: studies on the regulation of DNA synthesis and imtoris. *Nature, Lond.*, **225**, 159–64.

RAPER, K. B. and FENNELL, D. I. (1952). Stalk formation in *Dictyostelium*. *Bull. Torrey bot. Club*, **79**, 25–51.

RAVEN, C. P. (1963). Differentiation in mollusc eggs. In 'Cell Differentiation'. *Symp. Soc. exp. Biol.* no. 17, 274–84.

REES, H. and JONES, R. N. (1972). The origin of the wide species variation in nuclear DNA content. *Int. Rev. Cytol.*, **32**, 53–92.

RIGGS, A. (1951). The metamorphasis of hemoglobin in the bullfrog. *J. gen. Physiol.*, **35**, 23–40.

RIS, H. and KUBAI, D. F. (1970). Chromosome structure. *A. Rev. Genet.*, **4**, 263–94.

ROBBINS, W. J. (1957). Gibberellic acid and the reversal of adult *Hedera* to a juvenile state. *Am. J. Bot.*, **44**, 743–6.

ROBINSON, G. A., BUTCHER, R. W. and SUTHERLAND, E. W. (1971). *Cyclic AMP*. New York and London: Academic Press.

ROGOLSKY, N. and SLEPECKY, R. A. (1964). Elimination of a genetic determinant for sporulation of *Bacillus subtilis* with acraflavin. *Biochem. biophys. Res. Commun.*, **16**, 204–8.

ROTHSCHILD, M. (1965). The rabbit flea and hormones. *Endeavour*, **24**, 162–8.

RUTTER, W. J., KEMP, J. D., BRADSHAW, W. S., CLARK, W. R., RONZIO, R. A. and SANDERS, T. G. (1968). Regulation of specific protein synthesis in cytodifferentiation. *J. Cell Physiol.*, **72**, Suppl. I, 1–18.

RUTTER, W. J., PICTET, R. L. and MORRIS, P. W. (1973). Toward molecular mechanisms of developmental processes. *A. Rev. Biochem.*, **42**, 601–46.

SAUER, H. W. (1973). Differentiation in *Physarum polycephalum*. In 'Microbial Differentiation'. *Soc. for Gen. Microbiol. Symp.*, **23**, 375–405.

SAUNDERS, G. C. and WILDER, M. (1971). Repetitive maturation cycles in a cultured mouse myeloma. *J. Cell Biol.*, **51**, 344–8.

SAXEN, L. (1971). Inductive interactions in kidney development. In 'Control Mechanisms of Growth and Differentiation'. *Symp. Soc. exp. Biol.* no. 25, 207–21.

SCHARRER, E. (1965). The final common path in neuroendocrine integration. *Archs Anat. microsc. Morph. exp.*, **54**, 359–70.

SCHARRER, E. and BROWN, S. (1962). Electron-microscopic studies of neurosecretory cells in *Lumbricus terrestris*. *Mem. Soc. Endocr.*, **12**, 103–8.

SCHIMKE, R. T. (1969). On the roles of synthesis and degradation in regulation of enzyme levels in mammalian tissues. In *Current Topics in Cellular Regulation*, Vol. I, 77–124.

SCHIMKE, R. T. (1973). Control of enzyme levels in mammalian tissues. *Adv. Enzymol.*, **37**, 135–87.

SCHNEIDER, R. G. and HAGGARD, M. E. (1955). Sickling: a quantitatively delayed genetic character. *Proc. Soc. exp. Biol. Med.*, **89**, 196–9.

SCHÖTZ, F. (1970). Effects of the disharmony between genome and plastome on the differentiation of the thylakoid system in Oenothera. In 'Control of Organelle Development', *Symp. Soc. exp. Biol.* no. 24, 39–54.

SCHWEIGER, H. G. (1970). Synthesis of RNA in *Acetabularia*. In *Control of Organelle Development, Symp. Soc. exp. Biol.* no. 24, 327–44.

SHAW, C. R. (1969). Isozymes: classification, frequency, and significance. *Int. Rev. Cytol.*, **25**, 297–332.

SHAW, L. M. and HUANG, R. C. (1970). A description of two procedures which avoid the use of extreme pH conditions for the resolution of components isolated from chromatins prepared from pig cerebellar and pituitary nuclei. *Biochemistry, Easton*, **9**, 4530–42.

SHOPE, R. E. and HURST, E. W. (1933). Infectious papillomatosis of rabbits. *J. exp. Med.*, **58**, 617–24.

SMITH, G. M. (1955). *Cryptogamic Botany*, Vol. I, 2nd ed. New York: McGraw-Hill.

SOLTER, D. (1975). Early mammalian development. Embryo-derived teratoma: a model system in developmental and tumour biology. In *2nd Symp. Soc. Devl Biol*.

SONNEBORN, T. M. (1963). Does preformed cell structure play an essential role in cell heredity? In *The Nature of Biological Diversity, The University of Michigan Institute of Science and Technology Series* (ed. J. M. Allen), 165–221. New York: McGraw-Hill.

SOUTHERN, E. M. (1970). Base sequence and evolution of guinea-pig α-satellite DNA. *Nature, Lond.*, **227**, 794–8.

SPEMANN, H. (1938). *Embryonic Development and Induction*. New Haven, Connecticut: Yale Univ. Press.

SPOHR, G., GRANBOULAN, N., MOREL, C. and SCHERRER, K. (1970). Messenger RNA in HeLa cells. Investigation of free and polyribosome-bound cytoplasmic messenger ribonucleoprotein particles by kinetic labelling and electron microscopy. *Eur. J. Biochem.*, **17**, 296–318.

STAHL, F. W. (1964). *The Mechanics of Inheritance*. Englewood Cliffs, New Jersey: Prentice-Hall Inc.

STANIER, R. Y., DOUDOROFF, M. and ADELBERG, E. A. (1971). *General Microbiology.* London: Macmillan.

STEINBERG, M. S. (1970). Does differential adhesion govern self-assembly processes in histogenesis? Equilibrium configurations and the emergence of a hierarchy among populations of enbryonic cells. *J. exp. Zool.*, **173**, 395–433.

STEINER, A. L. (1970). The measurement of cyclic nucleotides by radioimmunoassay. *J. clin. Invest.*, **49**, 93a.

STEVENS, L. C. (1970). The development of transplantable teratocarcinomas from intratesticular grafts of pre- and post-implantation mouse embryos. *Devl Biol.*, **21**, 364–82.

STEVENS, L. C. and LITTLE, C. C. (1954). Spontaneous testicular tumours in an inbred strain of mice. *Proc. natn. Acad. Sci. U.S.A.*, **40**, 1080–7.

STEWARD, F. C. (1968). *Growth and organisation in plants.* Reading, Mass: Addison-Wesley Pub. Co.

STEWARD, F. C., MAPES, M. O. and SMITH, J. (1958). Growth and organized development of cultured cells. I. Growth and division of freely suspended cells. *Am. J. Bot.*, **45**, 693–703.

STREHLER, B. L. (1962). *Time, Cells and Aging.* New York: Academic Press.

STRICKBERGER, M. W. (1971). *Genetics.* London: Collier-Macmillan Ltd.

SUEOKA, N. and KANO-SUEOKA, T. (1970). Transfer RNA and cell differentiation. *Prog. Nucleic Acid Res.*, **10**, 23–55.

SUGINO, A., HIROSE, S. and OKAYAKI, R. (1972). RNA-linked nascent DNA fragments in *Escherichia coli. Proc. natn. Acad. Sci. U.S.A.*, **69**, 1863–7.

SUTHERLAND, E. W. and RALL, T. W. (1958). Fractionation and characterization of a cyclic adenine ribonucleotide formed by tissue particles. *J. biol. Chem.*, **232**, 1077–91.

SWANSON, C. P., MERZ, T. and YOUNG, W. J. (1967). *Cytogenetics.* Prentice-Hall Foundations of Modern Genetics Series. Englewood Cliffs, New Jersey: Prentice-Hall.

TAURO, P., SCHWEIZER, E., EPSTEIN, R. and HALVORSON, H. O. (1969). Synthesis of macromolecules during the cell cycle in yeast. In *The cell Cycle, Gene-Enzyme Interactions* (ed. G. M. Padilla, G. L. Whitson and I. L. Cameron), 101–18. New York and London: Academic Press.

THOMAS, C. A. (1970). The theory of the master gene. In *The Neurosciences second study program* (ed. F. O. Schmitt), 973–98. New York: Rockefeller University Press.

THOMAS, C. A., HAMKALO, B. A., MISRA, D. N. and LEE, C. S. (1970). Cyclization of eucaryotic deoxyribonucleic acid fragments. *J. molec. Biol.*, **51**, 621–32.

THOMPSON, L. R. and MCCARTHY, B. J. (1967). Stimulation of nuclear DNA synthesis by cytoplasmic extracts *in vitro. Biochem. biophys. Res. Commun.*, **30**, 166–72.

TODD, D., LAI, M. C. S., BEAVEN, G. H. and HUEHNS, E. R. (1970). The abnormal haemoglobins in homozygous alpha-thalassemia. *Br. J. Haemat.*, **19**, 27–31.

TOMKINS, C. (1968). In *Regulatory mechanisms for protein synthesis in mammalian cells* (ed. A. San Pietro, M. R. Lamborg, F. T. Keverey). New York: Academic Press.

TOMKINS, G. M., GELEHRTER, T. D., NARTIN, D., SAMUELS, H. H. and THOMPSON, E. B. (1966). Control of specific gene expression in higher organisms. *Science, N.Y.*, **166**, 1474–80.

TOMKINS, G. M., LEVINSON, B. B., BAXTER, J. D. and DETHLEKSEN, L. (1972). Further evidence for post-transcriptional control of inducible tyrosine aminotransferase synthesis in cultured hepatoma cells. *Nature, Lond. (New Biol.)*, **239**, 9–14.

TOWNES, P. L. and HOLTFRETER, J. (1955). Directed movements and selective adhesion of embryonic amphibian cells. *J. exp. Zool.*, **128**, 53–120.

TRUNKAUS, J. P. (1969). *Cells into Organs: the forces that shape the embryo.* Englewood Cliffs, New Jersey: Prentice-Hall.

TURKINGTON, R. W. (1968). Cation inhibition of DNA synthesis in mammary epithelial cells *in vitro. Experientia*, **24**, 226–8.

TURKINGTON, R. W. (1971). Hormonal regulation of cell proliferation and differentiation. In *Developmental Aspects of the Cell Cycle* (ed. I. L. Cameron, G. M. Padilla and A. M. Zimmerman), 315–55. New York and London: Academic Press.

ULLMANN, S. L. (1965). Epsilon granules in *Drosophila* pole cells and oocytes. *J. Embryol. exp. Morph.*, **13**, 73–81.

VAN WAGTENDONK, W. J., CLARK, J. A. D. and GODOY, G. A. (1963). The biological status of lambda and related particles in *Paramecium aurelia. Proc. natn. Acad. Sci. U.S.A.*, **50**, 835–8.

VASIL, V. and HILDEBRANDT, A. C. (1967). Further studies on the growth and differentiation of single, isolated cells of tobacco *in vitro. Planta*, **75**, 139–51.

WADDINGTON, C. H. (1956). *Principles of Embryology.* London: George Allen & Unwin.

WALKER, P. M. B. (1971). 'Repetitive' DNA in higher organisms. In *Progress in Biophysics and Molecular Biology*, Vol. 23 (ed. J. A. V. Butler and D. Noble), 145–90. Oxford: Pergamon Press.

WAREING, P. F. (1971). Some aspects of differentiation in plants. In *Control Mechanisms of Growth and Differentiation. Symp. Soc. exp. Biol.* no. 25, 323–44.

WATSON, J. D. (1970). *Molecular Biology of the Gene* (second edition). New York: W. A. Benjamin, Inc.

WATTS, R. L. (1971). Genes, chromosomes and molecular evolution. In *Molecular Evolution*, Vol. II. 'Biochemical Evolution and the Origin of Life' (ed. E. Schoffeniels), 14–42. Amsterdam: North-Holland Publishing Co.

WEBSTER, G. (1971). Morphogenesis and pattern formation in hydroids. *Biol. Rev.*, **46**, 1–46.

WEISOGEL, P. O. and BUTTOW, R. A. (1971). Control of the mitochondrial genome in *Saccharomyces cerevisiae. J. biol. Chem.*, **246**, 5113–9.

WHITEHOUSE, H. L. K. (1967). A cycloid model for the chromosome. *J. Cell Sci.*, **2**, 9–22.

WIENER, F., KLEIN, G. and HARRIS, H. (1973). The analysis of malignancy by cell fusion. IV. Hybrids between tumour cells and a malignant L cell derivative. *J. Cell Sci.*, **12**, 253–61.

WIGGLESWORTH, V. B. (1954). *The Physiology of Insect Metamorphosis.* Cambridge: University Press.

WILKIE, D. (1964). *The Cytoplasm in Heredity.* London: Methuen.

WILLIAMSON, R., DREWIENKIEWICZ, C. E. and PAUL, J. (1972). Globin messenger sequences in high molecular weight RNA from embryonic mouse liver. *Nature, Lond. (New Biol.)*, **241**, 66–8.

WILLMER, E. N. (1958). Further observations on the 'Metaplasia' of an amoeba, *Nae gleria gruberi. J. Embryol. exp. Morph.*, **6**, 187–214.

WILSON, E. B. (1925). *The cell in Development and Heredity.* New York: Macmillan.

WIPF, L. and COOPER, D. C. (1938). Chromosome numbers in nodules and roots of red clover, common vetch and garden pea. *Proc. natn. Acad. Sci. U.S.A.*, **24**, 87–91.

WITTWER, S. H., BUKOVAC, M. J., SELL, H. M. and WELLER, L. E. (1957). Some effects of gibberellin on flowering and fruit setting. *Pl. Physiol., Lancaster*, **32**, 39–41.
WOLPERT, L. (1971). Positional information and pattern formation. In *Current Topics in Developmental Biology*, Vol. 6 (ed. A. A. Moscona and A. Monroy), 183–224. New York: Academic Press.
WOODWARD, D. O., EDWARDS, D. L. and FLAVELL, R. B. (1970). Nucleocytoplasmic interactions in the control of mitochondrial structure and function in *Neurospore*. In 'Control of Organelle Development'. *Symp. Soc. exp. Biol.* no. 24, 55–69.
WRIGHT, B. E., DAHLBERG, D. and WARD, C. (1968). Cell wall synthesis in *Dictyostelium discoideum*. A model system for the synthesis of alkali-insoluble cell wall glycogen during differentiation. *Archs Biochem. Biophys.*, **124**, 380–5.
YOUNGER, K. B., BANERJEE, S., KELLGHER, J. K., WINSTON, M. and MARGULIS, L. (1972). Evidence that the synchronized production of new basal bodies is not associated with DNA synthesis in *Stentor coeruleus*. *J. Cell Sci.*, **11**, 621–37.

Additional references

BALDWIN, J. P., BOSELEY, P. G., BRADBURY, E. M. and IBEL, K. (1975). The subunit structure of the eukaryotic chromosome. *Nature, Lond.*, **253**, 245–9.
BRETSCHER, M. S. and RAFF, M. C. (1975). Mammalian plasma membranes. *Nature, Lond.*, **258**, 43–9.
COX, R. F., HAINES, M. E. and EMTAGE, J. S. (1974). Quantitation of ovalbumin m RNA in hen and chick oviduct by hydridization to complementary DNA. *Eur. J. Biochem.*, **49**, 225–36.
GURDON, J. B. (1974). *The control of gene expression in animal development*. Oxford: Clarendon Press.
HUTCHINSON, C. A., NEWBOLD, J. L., POTTER, S. S. and EDGELL, M. M. (1974). Maternal inheritance of mammalian mitochondrial DNA. *Nature, Lond.*, **251**, 536–8.
KIVILAAKSO, E. and RYTOMAA, T. (1971). Erythrocyte chalone, a tissue-specific inhibitor of cell proliferation in the erythron. *Cell Tissue Kinet.*, **4**, 1–9.
MACLEAN, N. (1973). Suggested mechanism for increase in size of the genome. *Nature, Lond. (New Biol.)*, **246**, 205–6.
MACLEAN, N. (1976). *The Control of Gene Expression*. New York & London: Academic Press.
MACLEAN, N. and HILDER, V. A. (1976). Mechanisms of chromatin activation and repression. *Int. Rev. Cytol.* (in the press).
THORNLEY, A. L. and LAURENCE, E. B. (1975). The present state of biochemical research on chalones. *Int. J. Biochem.*, **6**, 313–20.
SPEIRS, J. and BIRNSTIEL, M. (1974). Arrangement of the 5.8 S RNA cistrons in the genome of *Xenopus laevis*. *J. molec. Biol.*, **87**, 237–56.

Index